DEFINITIONS AND DEFINABILITY: PHILOSOPHICAL PERSPECTIVES

VOLUME 216

DEFINITIONS AND DEFINABILITY: PHILOSOPHICAL PERSPECTIVES

Edited by

JAMES H. FETZER

Department of Philosophy
University of Minnesota, Duluth, U.S.A.

DAVID SHATZ

Department of Philosophy
Yeshiva University, New York, U.S.A.

and

GEORGE N. SCHLESINGER

Department of Philosophy
University of North Carolina, Chapel Hill, U.S.A.

KLUWER ACADEMIC PUBLISHERS

DORDRECHT / BOSTON / LONDON

Library of Congress Cataloging-in-Publication Data

Definitions and definability : philosophical perspectives / edited by
James H. Fetzer, David Shatz, and George Schlesinger.
p. cm. -- (Synthese library ; v. 216)
Includes index.
ISBN 0-7923-1046-2 (alk. paper)
1. Definition (Logic) I. Fetzer, James H., 1940- . II. Shatz,
David. III. Schlesinger, George N. IV. Series.
BC199.D4D43 1991
160--dc20 90-21983

ISBN 0-7923-1046-2

Published by Kluwer Academic Publishers,
P.O. Box 17, 3300 AA Dordrecht, The Netherlands.

Kluwer Academic Publishers incorporates the publishing
programmes of D. Reidel, Martinus Nijhoff,
Dr W. Junk and MTP Press.

Sold and distributed in the U.S.A. and Canada
by Kluwer Academic Publishers,
101 Philip Drive, Norwell, MA 02061, U.S.A.

In all other countries, sold and distributed
by Kluwer Academic Publishers Group,
P.O. Box 322, 3300 AH Dordrecht, The Netherlands.

Printed on acid-free paper

Printed in the Netherlands

To
George N. Schlesinger

TABLE OF CONTENTS

Preface ix

PROLOGUE

JAMES H. FETZER / Aspects of the Theory of Definition 3

PART I / PRELIMINARY CONSIDERATIONS

JAMES CARGILE / Real and Nominal Definitions 21

JAMES H. FETZER / Primitive Concepts: Habits, Conventions, and Laws 51

PART II / DEFINITIONAL DESIDERATA

ROY SORENSEN / Vagueness and the Desiderata for Definition 71

WILLIAM G. LYCAN / Definition in a Quinean World 111

PART III / FORMAL DEVELOPMENTS

VEIKKO RANTALA / Definitions and Definability 135

JAAKKO HINTIKKA / Towards a General Theory of Identifiability 161

PART IV / EPISTEMIC DIMENSIONS

DAVID SHATZ / Epistemic Terms and the Aims of Epistemology 187

STEVEN LUPER-FOY / Rational Definitions and Defining Rationality 203

PART V / SPECIALIZED CONCEPTIONS

ALAN BERGER / Idealized Definitions in Physics and Idealized Dispositions 223
GEORGE N. SCHLESINGER / Inverted Definitions and Their Uses 239

PART VI / DISCIPLINARY CONCEPTIONS

MICHAEL D. BAYLES / Definitions in Law 251
THOMAS V. MORRIS / Defining the Divine 269

EPILOGUE

MICHAEL LEVIN / Philosophical Analyses: An Explanation and Defense 287

Index of Names 311

Index of Subjects 315

PREFACE

This collection of original papers is intended to provide a representative cross-section of recent work on definitions and definability. It begins with a general introduction that provides a perspective for approaching issues of this kind, including an indication of some of the important aspects of these studies, and ends with a spirited defense of an approach toward philosophical analysis that many are likely to find appealing. The papers in between reflect a diverse assortment of methods and theories that offer strong evidence that a philosopher's conception of definition is closely related to his conception of methodology in philosophy. The significance of the accounts that are presented here, therefore, should not be interpreted too narrowly.

This anthology originated in the fertile mind of George Schlesinger, who carried the project forward in collaboration with David Shatz. All of those whose contributions are included here were invited to participate by them. At a later stage in the development of this work, James H. Fetzer was asked to organize these papers and to provide an introduction. Although it may seem unusual for us to dedicate this work to one of us, we cannot resist an opportunity of this kind to express our appreciation to our colleague for his friendship and for his contributions to philosophy. George is an original in his personality and in his creativity. Like other readers of this book, he is only going to discover this dedication the first time he opens its cover.

J. H. F. and D. S.

PROLOGUE

JAMES H. FETZER

ASPECTS OF THE THEORY OF DEFINITION

The nature of meaning and the definition of terms has preoccupied philosophers at least since the time of Socrates. The centrality of language with respect to the construction of arguments and the centrality of arguments to the advancement of philosophy makes the nature of meaning and the character of definitions of special importance in relation to philosophy. Indeed, these issues are sufficiently central that a philosopher's theory of definition comes close to qualifying as that philosopher's conception of methodology in philosophy. There can be little doubt of its importance for philosophy today.

Many schemes have been proposed for understanding the logical forms that definitions may assume, including Aristotle's influential conception of *proper definitions*. According to Aristotle, all definitions define a species by means of a genus and a differentia, where the differentia distinguishes the members of the more specific species from the members of the less specific genus. When the genus and the differentia reflect properties that things of the species have essentially (or "by their nature"), the result is a proper definition of things of that kind. Man is thus supposed to be a rational animal.

The least satisfactory aspect of the Aristotelian theory, no doubt, is the role that it assumes for *essential properties*, where the precise character of those properties that are essential has proven difficult to pin down. Notice, for example, that essential properties are properties that no member of the species could be without, since otherwise they would not be members of the species. If some men are tall and others are short, some are fat and others are thin, then these properties cannot qualify as essential to being human. But how can we know which things in the world can be taken to be human?

Without already knowing that some things in the world are and other things in the world are not instances of the species itself, in other words, it is impossible to eliminate any properties at all from those that might be essential to that species. Perhaps being short or being fat *are* properties that no member of this species can be without, in which case tall and thin things cannot be instances of things that are human.

J.H. Fetzer, D. Shatz and G. Schlesinger (eds.), *Definitions and Definability: Philosophical Perspectives*, 3–17.

Without already committing ourselves to some idea, however vague, of what it takes to be a thing of the kind under consideration, the Aristotelian approach cannot even be applied.

The problems that arise in attempting to define the kind of things that things are become less acute when a distinction is drawn between *real definitions*, on the one hand, and *nominal definitions*, on the other. For nominal definitions can be viewed as concerned with words as elements of language, while real definitions are concerned with the properties of things in the world. A methodology appropriate for nominal definition, after all, may imply the properties that things of that kind in the world must have, without presupposing the metaphysics of essential properties that is so problematic.

A contemporary theoretician of philosophical methodology who has articulated the difference at issue here is Carl G. Hempel. Hempel (1952) has distinguished between several species of definition, none of which generate the problems that Aristotelian definitions confront. *Nominal definition*, for example, involves the introduction of new words to have the same meaning as some old word, phrase, or expression. The acceptability of a definition of this kind, moreover, is strictly a matter of convention among the users of a language, where its principal contribution is for the purpose of abbreviation.

Hempel also distinguished between meaning analysis and empirical analysis, where the former concerns elements of language but the latter deals with things in the world. *Meaning analysis* thus reports the established usage of specific words within a specific language-using community during a specific temporal interval. *Empirical analysis* reports the properties which are possessed by things in the world. Meaning analysis assumes that there are no special problems involved in identifying members of the target language community, and empirical analysis assumes there are no special problems involved in identifying samples of the target thing-in-the-world kind.

The kind of definition with which we are all most familiar, no doubt, is that of meaning analysis, examples of which are reported in the dictionaries that are compiled for various languages and dialects used in different areas of the world. Examples of empirical analysis are not as familiar, if only because we seldom reflect on the role of theorizing in science with respect to its contribution to the definition of words. Nevertheless, as Hempel especially has emphasized, theory construction and concept formation are so intimately intertwined as to qualify as virtually two sides the very same coin.

By way of illustration, consider the case of gold. Humans have worked with gold for a very long time. The ancient Egyptians, for example, knew it to be a yellowish, malleable metal that could be readily worked into jewelry and other artifacts. What they did not know, however, was that gold has a melting point of 1063° C, that it has a boiling point of 2600° C, and that it has atomic number 79. Some of these discoveries may be viewed as trivial in relation to the use of a particular scale of measurement, but others, such as atomic number, surely are not. The discovery of its atomic number had to await the atomic theory of matter and the periodic table of the elements.

The Hempelian conception of empirical analysis thus improves upon the Aristotelian conception of real definition to the extent to which it abandons the metaphysics of essential properties. Because the color, the malleability, the melting and the boiling points of things that are gold are supposed to be related by law to those things that have atomic number 79 (as a function of the number of protons in the nucleus of atoms of that kind), however, there remains a residual problem, namely: providing an adequate account of the nature of *nomic properties* (which are possessed by natural law) as opposed to one of the nature of *essential properties* (which are possessed by nature).

In addition to these distinctions, Hempel sought to elaborate upon the nature of explication as this distinctive methodology was envisioned by Rudolf Carnap, especially, before him. An *explication* assumes the form of a recommendation (or a "proposal") as to how a certain word, phrase, or expression might best be understood in order to attain specific theoretical or philosophical objectives. It is especially applicable with respect to crucial notions, such as "theory", "law", and "explanation" within the context of the theory of science, where recommendations can be advanced concerning the conceptions that might be most illuminating within that intellectual domain.

There is no need to recount Hempel's investigation, which is a model of clarity of exposition. It can function here as a benchmark against which the more recent contributions that constitute the content of this collection can be measured. For if the authors and the works that are found within the pages of this book go beyond the analyses that Hempel previously proposed, there will be little doubt that progress has been attained in matters methodological. However this may turn out, there can be no doubt that the papers that are presented here offer a variety of interesting, original and insightful observations that ought to stimulate philosophers and non-philosophers alike.

1. PRELIMINARY CONSIDERATIONS

Anyone inclined to think that everything that could be said has already been said when it comes to definitions should be disabused of that belief by James Cargile's exploration of real and nominal definitions. Cargile takes the purpose of *nominal definitions* to be that of giving the meaning of linguistic entities (such as words and sentence), while that of *real definitions* is rather that of clarifying given meanings themselves (viewed as properties or propositions). He concedes that his position presupposes the existence of properties or propositions, which nominalists may find theoretically objectionable.

Cargile impugns the slogan, "Don't ask for the meaning, ask for the use", as absurd, "if we consider arbitrary uses that could be arranged, as in a secret code". From his perspective, almost any word(s) could mean almost any thing, under the right circumstances, however unusual those particular conditions might be. Although he acknowledges a special kind of "purely nominal" definition whose purpose is strictly abbreviatory, he denies that even this species of definition is therefore unproblematical. Few readers will be unable to appreciate the subtlety and complexity of the arguments he offers.

Although Cargile's inquiry is pursued from his Platonistic perspective, it raises several questions about the nature of "necessity" and "contingency" as applied to sentences within particular languages. A sentence that might qualify as a "necessary truth" within a language **L** viewed as an abstract object might be "merely contingent" were that language actually adopted by a human community. The point of this distinction, therefore, is conveyed by his contention that, unlike its abstract counterparts, "A language is a living thing which can survive some changes in meaning on the part of its words".

The point of departure for James H. Fetzer's contribution is the familiar recognition that every word can be defined within a language only on pain of definitional circularity or of definitional regress. His attention is focused upon the problem of accounting for the meaning of *primitive* (or undefined) words, where the meaning of every defined word is reducible to the meaning of some combination of primitives. The alternative hypotheses include that primitive meanings are determined by natural laws, by personal habits, or by social conventions. Here the role of dispositions appears fundamental.

Fetzer suggests that primitive concepts can be characterized by means of their causal role in influencing behavior. On the basis of Peirce's theory of signs, he proposes a Peircean theory of minds, according to which "minds" are semiotic (or "sign-using") systems. The meaning of a specific sign, such as the use of specific words on a specific occasion, for a person *z* consists in the totality of tendencies toward behavior for that person in the presence of that sign, relative to every possible context (where "contexts" consist of other motives, beliefs, and so on, that make a difference to his tendencies).

The position that is elaborated here thus contends that the fundamental unit of meaning for a language user *z* appears to be a function of that user's habits, tendencies, and dispositions, while the fundamental unit of meaning for a community of language users *Z* is a function of those users' shared habits, tendencies, and dispositions. The success of efforts to ensure that different members of any such community share the same linguistic habits, tendencies, and dispositions, however, appears to depend upon the existence of natural laws of various kinds, which are presupposed by social conventions.

2. DEFINITIONAL DESIDERATA

Roy Sorenson attacks the problem of vagueness in relation to definitions, suggesting that there is something about it that runs against a philosopher's grain: "We like our theories to be consistent, complete, general, simple, fruitful — and precise". His discussion ranges broadly over the views of a number of well-known figures — including Goodman, Lehrer, Quine, Unger, Lycan, van Fraassen, Armstrong, Dreyfus, Fodor, and others still — in an effort to clarify the exact relationship between borderline cases and acceptable definitions, which extends to considering whether vagueness deserves to be eradicated.

His essay abounds with fascinating illustrations of doctrines and beliefs about language and definitions that cannot withstand scrutiny. It sounds plausible to suppose that two vague expressions could not create a precise expression merely by being conjoined, yet "integer somewhat greater than 104" and "integer somewhat less than 106" combine to form a precise term. Sometimes we accept a definition for technical reasons, such as the case in which Sorenson awards a prize for "fattest kitten in the world", where the artificial conception of a kitten as a cat younger than six months serves its purpose well, even though, in other situations, it might not be satisfactory.

Definitions that deliberately eliminate borderline cases are referred to as "precisifying definitions", where exactly how precise a definition needs to be may depend upon the objectives at hand. He takes to task the "supervaluationist" position that tends to envision making definitions more precise as an (almost) inevitable consequence of scientific progress. But he also concedes the conservative influence of convention preservation, which attempts to minimize variation in meaning across time in an effort to reduce mutual misunderstanding: "All of this forces revisionary definers to think small".

William Lycan undertakes the arduous task of revisiting the analytic/synthetic distinction in the aftermath of Quine's influential critique. He believes that most philosophers accept Quine's rejection of that distinction even though they reject Quine's principal rationale for its rejection, which is the thesis of the indeterminacy of translation. He disagrees with Quine in thinking that propositions *can* have determinate meaning (in relation to sets of possible worlds), while agreeing with Quine in doubting that there are sentences in natural language that are true by virtue of their meaning.

He proceeds by exploring several "unpromising accounts", including the view that truth in virtue of meaning means what it says because sentences do have meaning and some meanings guarantee truth (which he associates with Russell and Moore); the view that some sentences are true by virtue of their unfalsifiability (which he associates with Logical Positivism in various of its guises); the view that some sentences are true by convention (which he thinks has been partially redeemed by David Lewis' more recent work); and last but not least the view that some sentences are true by stipulation.

While Lycan admits that Quine himself has been "startlingly respectful" of the idea that some sentences might be true by stipulation in the case of the explicit introduction of novel notations merely for the sake of abbreviation, the majority of his efforts are devoted to establishing that there are good reasons for doubting that even this account will do what it must do if analyticity is to survive, including the observation that "No sentence that has ever been put forward as a necessary truth by a philosopher seriously philosophizing is [or has ever been] the product of a stipulative definition."

3. FORMAL DEVELOPMENTS

In his impressive study of definitions and definability, Veikko Rantala brings formal methods to bear on the traditional distinction between nominal and real definitions with which we began. He considers classic work by Russell and Whitehead (1910), by Cohen and Nagel (1961), and by Robinson (1965), discovering certain limitations inherent in the approaches that they represent. Robinson, for example, distinguishes between definitions of these kinds intentionally, where nominal definitions are intended to assign meaning to words and real definitions are intended to report the nature of things.

Appealing to speech act theories elaborated by Austin (1962) and Searle (1979), Rantala contends that, under the proper performative conditions, if an utterance has the appropriate syntactical form, whether or not an act of definition has taken place does not depend upon the presence or absence of corresponding intentions. As a consequence, the syntactical form of an utterance tends to be the most important factor in determining whether or not a word has been defined, where formal methods can be employed to capture the specific forms that are most appropriate for various kinds of definitions.

Building on previous results of Beth and Tarski, Rantala systematically develops formal notions of definability in a theory and of definability in a model. This provides a basis for capturing the difference between "global" and "local" definability and for exploring the requirements of eliminability and of noncreativity for definitions in formal contexts. Rantala reflects upon the possibility of applying these concepts to natural languages, concluding that difficulties with the notions of meaning and synonymy in natural language render this problematical. He concludes with a discussion of conditions under which definitions can be fruitful and extend our knowledge.

Jaakko Hintikka supplies the foundations for a general theory of identifiability that is intended to parallel the general theory of definability with its classification of different types of definitions, their interrelations, and their criteria of application. The difference between "identifiability" and "definability" is that the former depends upon empirical information that the latter does not require. The theory of definability, in other words, can be pursued as a purely formal enterprise, but the theory of identifiability defines concepts by using a formal theory together with empirical results.

The motivation for developing a suitable theory of identifiability arises primarily from its importance for understanding concepts that occur within empirical science. As an illustration from the history of science, he cites the problem posed by Ernst Mach of whether the concept of mass is definable in classical mechanics. When properly understood, what Mach wanted was not an explicit definition of "mass" on the basis of some axiomatization of classical mechanics, but rather to know "whether these laws enable a physicist to determine in each and every case the value of a point mass by means of suitable observations". It concerned identifiability rather than definability.

Employing an interrogative model of inquiry, Hintikka proceeds to formulate game-theoretic rules for interrogative derivability, which are applied to capture the notions of explicit definability, piecewise definability, and implicit definability. These results are extended (utilizing a theorem of Beth) to develop corresponding notions of explicit identifiability, piecewise identifiability, and (strong and weak) implicit identifiability. The importance of these results is illustrated by several cases that reflect controversies in the philosophy of science regarding incommensurability claims for classical and relativistic physics and the general ideas of observability and measurability.

4. EPISTEMIC DIMENSIONS

David Shatz explores many of the moves that theoreticians of knowledge have made in response to the challenge of skepticism. His primary purpose is to examine issues of two kinds, namely: methodological issues concerning the relationship between the untutored intuitions of ordinary people, on the one hand, and evaluative issues concerning whether or not the definition of epistemic terms is an important activity, on the other. During the course of his inquiry, Shatz considers at least three different arguments: the possibility argument, the ordinary believer argument, and the discovery argument.

A reliabilist view that knowledge consists of beliefs that are the product of a reliable process, for example, could be supported by the argument that skeptics cannot show that beliefs are never produced by reliable processes, which continues to be possible. The contention could also be made that no account of knowledge can succeed unless it follows that the ordinary beliefs of ordinary believers frequently qualify as knowledge. Alternatively, some theoreticians incline toward the view that, once an adequate conception has been advanced, we must discover whether or not anyone possesses knowledge.

Shatz distinguishes "epistemic" skepticism from "dialectical" skepticism on the grounds that the former maintains that we possess no knowledge, while the latter maintains that we are not justified in believing that we do possess knowledge. He thus suggests that the possibility argument does not succeed because the problem of skepticism is dialectical; that there is no good reason to suppose that ordinary believers must turn out to possess knowledge; and that the discovery argument does not adequately reflect the full dimensions of the problem, since it presupposes the availability of an account that both skeptics and non-skeptics could agree upon for application to specific cases.

Steven Luper-Foy approaches these issues as problems in reducing cognitive diversity, where philosophers of a certain "analytic" persuasion think that principles and procedures might be discovered which would enable us to distinguish acceptable cognitive states from unacceptable ones. One of the processes employed in pursuit of this approach, the "reflective equilibrium" test advanced by John Rawls, however, has been assailed by Stephen Stich, especially because it takes for granted that intuitions and attitudes that are displayed by ordinary believers provide relevant evidence for those views.

Beginning with Rawl's conception of the nature of reflective equilibrium in developing the concept of justice, Luper-Foy extends its scope to encompass inquiries in epistemology and in analytic philosophy generally. As he remarks, the defense of views on the nature of justice involves harmonizing a particular definition with a multitude of judgments about particular cases, which represent a form of "conceptual coherentism". Precisely similar considerations apply with respect to the concept of knowledge, where a suitable defense of definitions of the nature of knowledge involves a similar process.

Luper-Foy focuses attention on diverse views about the objective of epistemic justifications, including that of attaining true beliefs (which Laurence Bonjour and Alvin Goldman, among others, have endorsed) and that of maximizing true beliefs and minimizing false beliefs (which Richard Foley, especially, has proposed). Contending that some beliefs are more important than others even when they are false, Luper-Foy suggests that a more adequate conception of the aim of inquiry is provided by a satisficing approach as opposed to a maximizing one, where whether or not we understand what is the case well enough tends to depend upon our aims, goals, and objectives.

5. SPECIALIZED CONCEPTIONS

Although it may not be obvious from his title, Alan Berger's contribution concerns the use of idealized notions in the naturalistic philosophy of W. V. O. Quine. In particular, he focuses upon Quine's dispositional conception of logic, where logical truths are supposed to be ones to which speakers would assent immediately "under the appropriate prompting conditions". Thus, if a speaker were to fail to assent to a sentence that expresses a logical truth, then this impropriety could only be accounted for by the absence of those conditions. On Quine's account, these relations are not subject to variation.

A traditional linguist, by comparison, could treat a speaker's assents or dissents merely as useful evidence in drawing inferences about what that person does or does not believe. From this point of view, a speaker might never assent to a logical truth under any conditions at all. The defense of Quine's position must therefore appeal to the idea that the prompting conditions appropriate for eliciting the proper response should tacitly correct that behavior and change it into what it ought to be. But then "Quine is not interested in actual speech behavior except to change it into the verbal behavior that speakers *ought* to exhibit" on pain of the commission of various fallacies.

Berger thus suggests that Quine's account is fundamentally flawed by its implicit appeal to idealized speech dispositions that are *idealized toward a norm*, because there should be no place for normative ingredients within a naturalistic conception of logic. He further contends that, even though this use of idealized notions concerning language may appear to parallel the use of idealized notions concerning phenomena in physics, the differences that distinguish them are sufficient to destroy any analogy. Thus, language for Quine cannot be reduced to inductive regularities concerning actual usage within a community but must preserve rules that govern its proper usage.

The conception of inverted definitions that George Schlesinger examines appears to possess significance for definitions generally. On the basis of the notion of conceptual priority, he proposes a fundamental desideratum that acceptable definitions ought to satisfy, namely: that the definiendum of an acceptable definition must not be conceptually prior to its definiens. Those who may be skeptical of the utility of this approach ought to find that some of his examples — Jaegwon Kim's conception of events, Donald Davidson's conception of event identity, and W. V. O. Quine's conception of understanding — are surprisingly persuasive. Without doubt, he is not attacking straw men.

In relation to Kim's conception of events as exemplifications of properties by objects at times, Schlesinger observes that a universe might have events but no objects among its constituents. Events thus appear to be conceptually prior to objects and thus should not be defined in terms of them. Moreover, if acausal universes are possible, where, for example, every event occurs at random (spontaneously), an event *x* could still be identical with an event *y* yet they could not be said to have the same causes or the same effects. And Quine's conception of understanding, according to which "a man understands a sentence in so far as he knows its truth conditions", appears to imply satisfaction of a condition of belief which already presupposes understanding.

While Schlesinger rejects these three accounts because they violate this fundamental desideratum, most of his efforts are directed toward an evaluation of certain principles of deontic logic. On the basis of the assumption that an act *A* is morally obligatory in a situation of kind *B* if and only if the probability that an individual *z* picked at random from the set of righteous individuals who have ever found themselves under conditions *B* would do *A* is greater than 1/2, he offers a sustained critique that reveals apparently unacceptable consequences that follow from certain deontic axioms endorsed by Lewis and by Hansson, which is likely to provoke further discussion.

6. DISCIPLINARY CONCEPTIONS

Anyone who thinks that definitions only matter in philosophy ought to be enlightened by Michael Bayles' discussion of the role of definitions within the law. "We are not interested in whether the animal in question is a bird or not in fact, but whether it is one in law" indicates that all may not be as it seems to the untutored. As he emphasizes here, without knowing the purpose for which a definition is intended, its adequacy may be difficult, if not impossible, to assess. Deliberately shooting someone who is already dead in an effort to kill them might not qualify as attempted murder under the law, but in ethics there would be little doubt of moral culpability.

Surveying various attitudes that might be adopted toward definitions, where "realists" are those who think that definitions should capture properties that exist in the world, Bayles observes that "realism obviously does not require (logically, conceptually, practically, morally, or legally) that one provide realist definitions". Whether or not tomatoes ought to be classified as vegetables or as fruit for purposes of law, for example, should not rest upon the best scientific theories of the time but rather upon its legal consequences: "I, for one, am not at all disturbed by a court treating tomatoes as vegetables for the purpose of an import duty". Contexts make a difference.

Those who have ever doubted whether definitions matter should think about the Supreme Court's decision on abortion, which, as Bayles reports, required the Court to decide, not whether a fetus is a person for the purpose of ethics, but whether a fetus is a person for constitutional purposes. The purpose of definitions within the law, he suggests, is the development of a justifiable legal system. Those who imagine that this conception lets every judge decide for himself what words should mean (thereby inviting chaos in legal language) overlook the evident response that a system that permitted that result would not be justifiable (even on that ground alone).

The difficulties confronting definition in the law, however, are nothing compared to those confounding the definition of "God", as Thomas V. Morris explains. Traditional conceptions of God, after all, have viewed God as beyond comprehension: "It is difficult to conceive God, but to define him in words is an impossibility". For if God *is* the ultimate profound mystery, the prospects for his successful definition appear to be quite faint, indeed. Even the idea of an omnipotent, omniscient, and benevolent creator of the world, after all, makes God appear to be very different from familiar things.

Yet as Morris remarks, there are enormous differences between contending that God is not completely like anything in the world, on the one hand, and contending that God is completely unlike anything in the world, on the other. The distinction is the same as that between claiming that God does not share *all* properties with anything else and that God does not share *any* properties with anything else, which would appear to be impossible. Moreover, from another point of view, the problem appears not to be an absence of conceptions of God but an overabundance (including pantheism, polytheism, and monotheism), which raises the problem of choosing between them.

Morris attempts to resolve these differences by drawing upon two traditions in theology. The first, "Perfect Being Theology", tends to ascribe the properties of personal agency, perfect power, perfect knowledge, and perfect goodness to the deity. The second, "Creation Theology", defines "God" as the creator of the world. But, to avoid certain modal entanglements, it turns out to be preferable to envision God instead as the creator of whatever might exist apart from himself. Indeed, in order for God to be viewed as "the greatest possible being", he should be thought of as what Morris calls "an absolute creator", which forges a link between the two traditions.

7. PHILOSOPHICAL ANALYSES

Michael Levin contends that the differences between definitions and the analyses that philosophers provide run rather deep, partially because "good philosophical definitions have the ring of empirical discoveries or illuminating theories". Beneath the surface of qualification complicated by qualification and counterexample compounded by counterexample, he believes that a coherent and unified conception of language and analysis waits to be discovered. Assuming a "broadly behavioristic" conception of language, where idiolects are families of associations between utterances and contextual cues, semantic competence is nevertheless not innate but the product of learning.

Levin suggests that there really is "no generally accepted analytic characterization of philosophy", yet he denies that this is a case of the kind that Hilary Putnam has proposed of reference without meaning. He tackles the problem of vagueness, endorsing an account that counts the cases in which a speaker will assent to or dissent from the ascription of a predicate. He thus denies that vagueness poses the kinds of problem that Sorensen considers. Against this Quinean background, Levin suggests "that philosophical analyses are descriptions of verbal cues", which are attempts to uncover causes of verbal behavior of which speakers themselves are frequently unaware.

Among the striking features of Levin's account is that the meanings of words are characteristically learned by children as a function of contexts: explicit definitions are seldom advanced. An implication of his analysis is that semantics should not be read off of psychology but psychology should be read off of semantics. Levin illustrates his approach relative to Robert Nozick's conception of knowledge and even extends it to aesthetic criticism. By embracing counterfactuals and thought experiments to understand the role of cues, he supplies an account that many are likely to find appealing.

There are some surprising links between various positions. Cargile, the Platonist, for example, endorses the mind-dependence of language, which is asserted as the thesis that languages that express no properties to any mind cannot exist. Fetzer, the pragmatist, contends, with Peirce, that language involves a triadic relationship between words, what they stand for, and word users, where those things that can use language must also possess mentality. Sorensen suggests that the need for precision in definition tends to vary with the purpose at hand, which harmonizes with Fetzer's conception of beliefs as true when they are appropriate (sufficiently precise?) to guide our actions.

Analogously, the studies by Lycan, Rantala, and Hintikka, in very different ways, tend to suggest that formal models can be used to shed light on various aspects of natural language. Rantala attempts to illuminate certain pragmatical elements through speech act theory, while Hintikka explores the nature of empirical inquiry using his interrogative model. Even Lycan endorses the ascription of determinate meaning to propositions by means of possible worlds semantics. Each of their efforts thus offers further evidence that at least some of the distinctions that can be drawn by means of formal methods hold promise for clarifying our understanding of natural language.

With respect to the method of reflective equilibrium, Shatz tends to be a skeptic and Luper-Foy an advocate. While Shatz concludes that definitions of epistemic terms cannot provide an adequate response to dialectical skepticism, Luper-Foy rejects the maximization of truth and the minimization of falsity for a satisficing conception, which also seems to harmonize with the conception that beliefs are true (true enough?) when they are appropriate to guide our actions. And Berger's penetrating criticism of Quine's idealized dispositions provides further evidence that a dispositional conception of language that might succeed must be different in kind from Quine's account.

Schlesinger's notion of inverted definitions, of course, affords a foundation for evaluating definitional recommendations across a broad spectrum of disciplines extending far beyond the boundaries of philosophy as a domain. Bayles' analysis of definitions within the law, moreover, suggests that sometimes the most appropriate definition might be scientifically inappropriate, because the judicial function is to decide particular cases in accord with law. Perhaps most surprising, Morris' reconciliation of two traditions in theology strikingly resembles the combination of a definition with an identification.

Levin's explanation and defense of philosophical analyses, finally, bears certain affinities to Hempel's conception of methodology in matters of definition. Whether or not his Quinean conception of language can ultimately be sustained, his approach supports theses about dispositions and the conditions of adequacy for definitions that contrast with those of other authors in this volume. Anyone who is inclined to believe that behavioristic conceptions have had their day may want to reconsider after studying his position. Significantly, it once again appears as though concept formation and theory construction go hand in hand, in philosophy as much as in empirical science.

Department of Philosophy
University of Minnesota, Duluth

REFERENCES

Austin, J. L. (1970), 'Performative Utterances', in J. L. Austin, *Philosophical Papers* (Oxford, UK: Oxford University Press), pp. 233—252.

Cohen, M. and E. Nagel (1961), *An Introduction to Logic and Scientific Method* (London, UK: Routledge & Kegan Paul).

Hempel, C. G. (1952), *Fundamentals of Concept Formation in Empirical Science* (Chicago, IL: University of Chicago Press).

Robinson, R. (1965), *Definition* (Oxford, UK: The Clarendon Press).

Searle, J. (1973), *Expression and Meaning* (Cambridge, UK: Cambridge University Press).

Whitehead, A. N. and B. Russel (1910), *Principia Mathematica to *56* (Cambridge, UK: Cambridge University Press, 1973).

PART I

PRELIMINARY CONSIDERATIONS

JAMES CARGILE

REAL AND NOMINAL DEFINITIONS

Defining is either giving the meaning of words or other linguistic things, which is called "nominal definition" or clarifying a given meaning, which is a proposition or property or relation, by showing it to be a compound of other propositions or properties or relations, which is called "real definition".

The claim that the word "bachelor" means an unmarried male is a nominal definition and the thing defined is the word "bachelor". The definition gives the meaning of the thing defined, and the meaning is the property it expresses, the sense of the term.

That this nominal definition is a claim rather than a stipulation makes no difference to its logical status. It might be complained that there is more to being a bachelor than being an unmarried male, that the term is not applied to widowers, divorcees, babies or life-long human vegetables. In addressing this complaint, there has to be some specification of what uses of the term "bachelor" are in question, if there is to be a determinate question. Some may apply the term to divorcees, others may withhold it, more or less deliberately. To say "Bachelor' applies to divorcees" may be a claim about certain specific uses, or a claim about what the majority does. This latter sort of claim needs specification of standards for being a majority, and its basis must be the individual cases where someone succeeds in applying the term (whether truly or falsely) with a certain meaning.

This might seem to be a matter for philosophical disagreement. That the basic case is the use of a term with a certain meaning might seem to go against the spirit of the popular slogan, "Don't ask for the meaning, ask for the use." To me, this slogan is absurd if we consider arbitrary uses that could be arranged, as in a secret code. I would say that we should be interested only in uses made in accordance with the meaning. But the non-arbitrary stretchings involving such a confusing notion as metaphor suggest that use has a role that contributes to determining meaning, rather than its being true simply that meaning determines correct use. This can be a paralyzingly complex matter. But is should not count against the idea that if we do choose to speak of a term's

J.H. Fetzer, D. Shatz and G. Schlesinger (eds.), *Definitions and Definability: Philosophical Perspectives*, 21–50.

meaning, the property it expresses, we must be clear whether we are talking about specific occasions, or an average of some sort.

The word "bachelor" means an unmarried male or expresses the property of being an unmarried male to an agent on some occasion. If there is no such agent or occasion, then the claim as to what the word means is false. This holds even where there is a stipulation. I may say that it is unclear what people generally mean by "bachelor", so I am introducing "sachelor" to mean an unmarried male, regardless of previous marital history or condition, etc. Some will say this is "performatory", that I do not report my introducing the term, I just *do* introduce it, etc. It is still the same as a claim, and it may be false. I do not qualify as introducing "sachelor" to mean an unmarried male unless, among other things, "sachelor" does in fact mean this, express this property to someone.

This sort of nominal definition thus involves the idea of giving the sense, connotation, or meaning of the term or sentence defined. This meaning of "meaning" is an old one, presented in various ways by different philosophers, but widely criticized nowadays. It is said that "semantics" or "cognitive science" have no need or even any good use, for such a notion of meaning or content. Other uses of "meaning" are proposed. Perhaps these are interesting uses, and "cognitive science" or "semantics" may be so defined as to exclusively concern them. I will merely assume that there is reason to be interested in what is said, the proposition, or said of something, the property. In my terms, the meaning of "It is raining", for example, changes from one use to another, and is the proposition expressed at each particular use. Again, the property expressed by ". . . is here" changes. The variety of other uses of "meaning" are not our concern.

Even for holders of connotation-denotation accounts of meaning, it is not common to speak of defining a sentence or proposition. Subtle reasons might be offered, but I will merely say that it is as important to give or clarify the meaning of a sentence or proposition as it is for a term. Some may say that the meaning of a sentence can be calculated from the meanings of its constituent terms. To me, this is at the least not a simple procedure and the pursuit of it should not stop us from giving or clarifying the meaning more directly.

The claim that the word "bachelor" applies to all and only adult males may be made by way of offering a definition. But this is a misleading way of speaking. Both the claim that "bachelor" means an unmarried male and the claim that "bachelor" applies to all and only unmarried males are if true (and the latter is not true, incidentally) merely contingent facts. But the former is the fact that the word expresses the property of being an unmarried male. The latter fact does not entail the former. That the phrase "living passenger pigeon in 1910" applies to all and only birds in cage *A* does not entail that that phrase expresses the property of being a bird in cage *A*.

Note also that the claim that the word "bachelor" expresses the property of being a bachelor, that is, means being a bachelor, is a contingent factual claim meeting the requirements so far implied for being a nominal definition, but is not a nominal definition. This is tricky to explain. That a bachelor is a bachelor is not a real definition because it fails to represent a given property as a compound in a different way than first introduced. The case for the failed nominal definition is related.

Another caution relates to the style of reference to a property. Suppose that Alice is very impressed by the fact that Sam is a bachelor. We might say that the property of being a bachelor is the property of Sam's which most impressed Alice. This identity licenses the conclusion that the word "bachelor" expresses the property of Sam's which most impressed Alice. This is not a definition and it does not license the conclusion that the sentence, "A bachelor has the property of Sam's which most impressed Alice", expresses a necessary truth. It does license the conclusion that that sentence expresses a truth, but not a necessary one.

The word "bachelor" does express (we assume for example) the property of Sam's which most impressed Alice. But it does not express the property of being a thing which has the property of Sam's which most impressed Alice. On the other hand, it does express, not only the property of being an unmarried male, but also the property of being a thing with the property of being an unmarried male. This suggests a test for when an identification of a property expressed is a definition and when it is not. Whether or not the test works in general, caution is needed on this point.

What I call the nominal definition is very different logically from the real difinition, but both involve reference to properties or propositions. This might be called unfair to the nominalist, who wants to confine himself to talk about language. There is another sort of definition, which we may call *purely* nominal, the "abbreviative" definition, such as one telling us that for "unmarried male", the word "bachelor" is an acceptable substitute, as a rule of language. The purely nominal definition does not purport to give the meaning of "bachelor" but merely claims (whether stipulating or reporting) that it is replaceable by "unmarried male". Here "replaceable" is problematic, for example, in "Sam never calls anyone a bachelor, he always restricts himself to calling someone an unmarried male." Unless it is explained why "bachelor" and "unmarried male" cannot be interchanged in that example, without changing the truth value, the purely nominal definition of "bachelor" will be a false claim.

Furthermore, the word "bachelor" applies truly to certain college graduates and certain seals, in spite of their not being unmarried men. With the nominal definition, we can say that we are giving just one meaning of the word, and for that meaning, the word applies to all and only unmarried males. But the motives behind the purely nominal definition would require finding some other way of putting this, if there is one. And that is important, because without some such qualification, the purely nominal definition is just a false claim.

That both the nominal definition, which still employs the idea of a property, relation or proposition, and the purely nominal definition, which tries, at least, to avoid it, represent contingent claims, is widely denied or ignored. For example, in discussisng Tarski's theory of truth. Scott Soames holds that Tarski's theory calls for treating languages as abstract objects

> ... bearing their semantic properties essentially .... There is no possibility that expressions of a language might have denoted something other than what they do denote; or that the sentences of a language might have had different truth conditions. Any variation in semantic properties (across worlds) is a variation in languages. Thus semantic properties aren't contingent on anything, let alone speaker behavior. [Soames (1984), p. 425]

While this is in terms of denotation and "truth conditions" rather than sense, it is essentially the idea that, at least for a language taken as an "abstract object", the nominal definition of a term of language **L** defines the term as a term of **L** so that the property expressed by the term as a term of **L** is necessarily expressed by that term as a term of **L**.

It is perfectly possible to stipulatively define a language **L** as one in which, among other things, the term *t* expresses the property *F*. This could be a stipulative definition of the term "language **L**", or it could even be a real definition of a language of some specified sort. But it is similarly possible to define the property of being a rational root of three. It is no fault at all of the latter definition that nothing satisfies it, that there is no rational root of three. It is the definition that allows us to prove that it cannot be satisfied. Similarly, the definability of a language does not guarantee its existence. (And of course if I say there is no such language or no such root, I do not mean there is a non-existent one, but that there does not exist such a one.)

Defining a language **L** in such a way that it follows from the definition that in **L**, the sentence *S* expresses the proposition that *P*, or the term *t* expresses the property of being *F* does not make it true that in **L**, *S* expresses the proposition that *P* or *t* expresses the property of being *F*, any more than defining a love system *L* so that it follows from the definition that in *L*, Sam loves Penny makes it true that in *L*, Sam loves Penny. It merely makes Sam's loving Penny a necessary condition of the existence of such a system, or *S* or *t* expressing the required sense a necessary condition of the existence of the language, and no language exists necessarily. No matter how much you treat sentences or words as "types" and thus "abstract objects" (a questionable practice allowed here for purpose of argument), expressing a proposition or property is, like love, a relation requiring certain responses of people, with all their ineliminable contingency. As a platonist, I hold that the existence of propositions and properties is independent of people. But the existence of language expressing those properties is another matter.

It is not a necessary truth that the word "bachelor", as a word of English, means an unmarried male. The word "bachelor" does not have to mean an unmarried male in order to qualify as a word of English. Thus we have a double contingency. The existence of the English language is not necessary, and even given the existence of English, the role played in it by the word "bachelor" is not necessary. Could we, then, define a language **L** so that, in order to qualify as a word of **L**, "bachelor" would have to mean an unmarried male? If so, it would seem to follow that it is necessarily true that as a word of language **L** as an "abstract object", "bachelor" means an unmarried man, contrary to my insistence on contingency.

The argument would be as follows: the phrase "the word 'bachelor' as a word of **L**" means a word which means being an unmarried male. That nominal definition licenses the claim that it is necessarily true that the word "bachelor" as a word of **L** means being an unmarried male.

And this means that "bachelor" meaning unmarried male as a word of **L** is independent of such things as speaker behavior.

My answer is that this stipulative nominal definition makes the defined phrase "the word 'bachelor' as a word of **L**" into a fused expression. We could just as well have said that "alpha" means a word which means being an unmarried male. We may allow, for purpose of discussion, that this gives us the result that it is necessary that an alpha is a word meaning being an unmarried male, but this is nothing at all about the word "bachelor", not even as a word of **L**. When we drop the quotation we would get:

(1) The word "bachelor" as a word of **L** is a word that means being an unmarried male.

If this all succeeded, then (1) would be a necessary truth, but dropping the quotes to derive it would not warrant dropping the fusion. The subject of (1) would *not* be the word "bachelor" as a word of **L**. Rather, (1) would be a trivially necessary general claim about words which as words of **L** meant being an unmarried male, namely the claim that they are words which, as words of **L**, mean being an unmarried male. It would be difficult to get the original nominal definition to succeed, since the fusion is so awkward. But the idea that (1) could be derived as expressing the claim that the word "bachelor" as a word of **L** means being an unmarried male, and as a necessary truth, is going from the fusion to confusion.

Some philosophers will say that they are not speaking of the language spoken by such and such people at such and such time and place, but the language for which certain abstract rules hold. But even if the contingency is met that such a language comes to exist, those rules do not hold necessarily. If a language so defined has a word change meaning, then you can say it no longer qualifies as of **L**-type. But you can't say correctly that it no longer qualifies as the same language. A language is a living thing which can survive some changes in meaning on the part of its words.

It might be replied that while a language can change over its history, at a certain point it will have fixed meanings. This is like saying that while a man can change over his life, he is fixed at a given instant of it. The truth is that while a man cannot undergo change during an instant, he is nonetheless at the instant a thing that can change, and the same goes for any language. (Even a believer in the "Block Universe" should admit this, unless he is a Spinozist who rejects all contingency.)

The idea that we can define a language, and then the factual question is whether anyone speaks it, rather than whether there is such a language, might be fostered by the knowledge that there are in fact "dead" languages, that no one speaks now. So why not just extrapolate? If it can be a language not spoken now, what is so important about its having been spoken once? It was a problematic feature of a certain traditional style of definition that on its definition of a cat, a dead cat is not a cat, but a former cat. J. P. Downey has convinced me that this is wrong, and that it is perfectly possible logically that there should be a cat which was never alive — and not "a different breed of cat" — just a regular, say, Siamese, lying there full grown, dead and beginning to decay, having been created from nothing in that state. But it is not like this with language. There can be compounds — sentences or compound words — that no-one understands. But that an entire language should have no components that expressed any property to any mind is impossible. (It is ironical to a platonist that this doctrine of the mind-dependence of languages should qualify me, on certain contemporary definitions of the term, as an "anti-realist" about language.) Perhaps the exercise of defining a language will in some cases be done in such a way that the defining actually brings the language into existence by making terms serve to express properties for the definer. But this is not essential to defining a language and should not be taken for granted.

Here I recommend again this formula: a stipulative definition is always a factual claim. This is contrary to the idea that we couldn't be making a mistake in laying down a mere stipulation. The truth is that of course there is no falsity in succeeding in making a term express a certain property. But in attempting to do this, we may fail, or more importantly we may do something else which we misrepresent as stipulative defining.

Hilary Putnam has said, "Einstein, as we all know, changed the definition of 'kinetic energy'. That is to say, he replaced the law '$e = 1/2mv^2$' by a more complicated law." He goes on to say that, "It would clearly be a distortion of the situation to say that 'kinetic energy = $1/2mv^2$' was a definition, and that Einstein merely changed the definition."

This is a wise saying, but Putnam's view seems to be that 'kinetic energy = $1/2mv^2$' could begin as a stipulative definition and then be overthrown by the development of an alternative "conceptual system." He considers for purpose of discussion the possibility that, say, Newton and Leibniz disagreed as to whether to apply the term "energy" to "what we now do call 'energy' or what we call 'momentum'." We may imagine this settled by stipulation, and in that case "$e = 1/2mv^2$" would be true by stipulation, "*but only in a context which is defined by the fact that the only alternative principle is '$e = mv$'.*" He says:

> Quine has suggested that the distinction between truths by stipulation and truths by experiment is one which can be drawn only at the moving frontier of science. Conventionality is not a 'lingering trait' of the statements introduced as truths by stipulation. [Putnam (1966), p. 105]

It seems that Putnam would hold that the very same statement could go from being a truth by stipulation to being rejected as if it were a general law that had to be replaced. He says that before Einstein, physicists did not usually "suspect that the law '$e = 1/2mv^2$' might be wrong", but then this "law" was overthrown.

The idea that the overthrow of a truth, even one by stipulation, would be correct is to me uncongenial. It is abstractly possible that a scientist could claim that the phrase " 'kinetic energy' means $1/2mv^2$" could be at one time a true thing to utter, and later, express a false claim, due perhaps to the fact that Einstein had succeeded in appropriating the phrase 'kinetic energy' to express something different. But these would be two different claims. The sentence would have changed truth value, but only by expressing different propositions.

The "proposition" is not a notion acceptable to Putnam, but he would surely agree that the abstract possibility of introducing "kinetic energy" to mean $1/2mv^2$ does not correspond with the historical facts. I do not pretend to know the historical facts, but I believe that scientists originally thought about the energy an object acquires through motion. This they named "kinetic energy". They had the idea of the amount of work a body can accomplish through motion, from observing a body strike another and cause it to move. By sliding one block into another, they could calculate the distance the moving block could move the struck block as a function of its velocity and rest mass. They concluded that this observationally established law would hold generally. Einstein then disagreed, showing that the Newtonian law was not true. Those who represented this development as a change in definition were mistaken. Putnam seems to take the course of granting that the term "kinetic energy" was introduced to mean $1/2mv^2$ but that, in spite of this being so, the claim that kinetic energy is $1/2mv^2$ remained open to revision just like a general law.

My view, on the contrary, is that *if* it had been true that "kinetic energy" meant $1/2mv^2$ then in that situation, the sentence "kinetic energy is $1/2mv^2$" would have expressed a necessarily true proposition (though "'kinetic energy' means $1/2mv^2$" would still have expressed a contingently true proposition), and thus the content of that sentence would not be revisable, though of course the sentence could come to express some other proposition. The historical fact is that the proposition expressed was revisable, though Putnam would not speak in these terms — it was revisable in the sense appropriate for a proposition, namely, in turning out to have been false all along. But this in itself proves that the original "'kinetic energy' means $1/2mv^2$" was not a definition but a false claim.

This is not to say that someone might not have mistakenly regarded the formula as a definition and mistakenly regarded the claim that kinetic energy is $1/2mv^2$ as a necessary truth. These opinions, no matter how widespread, do not show anything about the necessity of the proposition or the truth of the meaning claim.

Obviously, opinion polls do have some role in finding out what words mean. But it is not always a straightforward one. People's responses to a term determine its meaning, but often in a complex, indirect way. It is a mistake to think that a stipulation can guarantee the required responses. Someone who sees the observed behavior of sliding blocks as confirming a formula for kinetic energy is confused if he then insists the formula reflects a stipulation. He is just not well aware of how his language functions.

This confusion might be lessened by carefully distinguishing between

(1) The words "kinetic energy" express the property of being half the rest mass times the square of the velocity,

and

(2) kinetic energy is half the rest mass times the square of the velocity.

In a language community in which sentence (1) expresses a true proposition, sentence (2) expresses a necessarily true proposition. Probably, sentence (2) never did express a necessary truth, and it certainly doesn't nowadays. But that means that (1) was probably never true and certainly isn't now. Conventionality and stipulativeness are not, to me, interesting properties and I do not care as to whether they linger or are confined to some "moving frontier". And perhaps the truth even of (1) would tend not to "linger." Still, it should be clear that while (1) is true, (2) expresses a necessary truth. If this seems implausible it is because in fact (1) never was true.

"True by virtue of meaning" is very misleading. The truth is that by virtue of sentence (1) being true (a counterfactual assumption for the purpose of discussion), sentence (2) would express a necesary truth. That is one use for "true by virtue of meaning". On the other hand, by virtue of its being true that "kinetic energy" expresses the property of being $1/2mv^2$, sentence (1) would be true. That is quite another way of being "true by virtue of meaning". Ignoring sentence (1) and focusing only on (2) (and in the form of a formula) makes it easier (perhaps) to tolerate the idea that the very same claim could go from being true by convention to being a false general law. Another safeguard here would be to note that if (2) were drawn from (1), it would be more clearly expressed by

(3) The property of being the kinetic energy of a body is the property of being one-half its rest mass times the square of its velocity.

To me it is clear that claim (3) would be either necessarily true or necessarily false, being not the claim that the two properties are just coextensive, but that they are one and the same property. This way of speaking may be rejected as platonistic and confusing. But I believe that on the contrary, it can be helpful in avoiding confusion.

Furthermore, it is wrong to say that

(4) From (1) it would follow that it is necessarily true that kinetic energy is half the rest mass times the square of the velocity.

What follows is that the proposition expressed by the sentence, "Kinetic energy is half the rest mass times the square of the velocity", is a necessarily true proposition. Since this is false, (1) is false. In general, a nominal definition will entail that a certain sentence expresses a necessarily true proposition. If we agree that a necessarily true proposition is trivially entailed by every proposition, then of course the nominal definition will entail that a certain proposition is necessary. But it does not produce or explain this necessity, as is brought out by the wrongness of (4).

Michael Dummett has said, "It is . . . plain that we may legitimately demand a certain consonance between the two aspects of the use of a given form of expression, and that the perception of the lack of such consonance is a major, and quite correct, ground for proposals to revise the sense of an expression or to reject it altogether" [Dummett (1977), p. 397].

These "two aspects of use" are rules for introducing a term, that is, rules for when the term applies and rules for "eliminating" the term, i.e., for what follows from its application. He goes on to say "if it is part of the established use of a certain form of sentence B" that it may be inferred from A and that C may be inferred from it, then "the presence of this form of sentence may be validly objected to if, in its absence, we should not be willing to recognize the legitimacy of the inference from A to C."

Later on, Dummett gives a specific example:

> A simple case would be that of a pejorative expression, e.g. 'Boche'. The condition for applying the term to someone is that he is of German nationality; the consequences of its application are that he is barbarous and more prone to cruelty than other Europeans. We should envisage the connections in both directions as sufficiently tight as to be involved in the very meaning of the word; neither could be severed without altering its meaning. Someone who rejects the word does so because he does not want to permit a transition from the grounds for applying the term to the consequences of doing so. The addition of the term 'Boche' to a language which did not previously contain it would be to produce a non-conservative extension, i.e. one in which certain statements which did not contain the term were inferable from other statements not containing it which were not previously inferable. [Dummett (1977), pp. 454—455]

These claims can of course make perfectly good sense if terms like "language", "inferable" and "meaning" have some meaning different than the one I give them. One possibility would be "language" as in "formalized language" (rather like "decoy duck") where a difference in axioms means a difference in language. But in the English language, people can dispute over any logical principle, such as noncontradiction, while speaking the same language. Of course people may speak the same language while meaning, in my sense, different things by many of their words. But this is not requred to allow the most fundamental logical disagreement.

There is some truth in the idea that by using the term 'Boche' I imply acceptance of certain negative stereotypes, and being ignorant of that is to be ignorant of the meaning, in some sense of 'meaning', of the term.

This is true generally of terms for racial and ethnic slurs. This is a valuable truth to articulate, and the matter is complex. However, this sort of "meaning" should never be confused with sense or content.

The account of the origin of the use of the word 'Boche' by British soldiers which is implicit in Dummett's remarks is, I believe, very far from a true account. But it is enough to ask whether it is possible, in terms of meaning as sense.

There is no problem with the proposal to use 'Boche' to mean a cruel and barbarous person. But this is not like baptizing a child 'Charles'. When the priest says, "I baptize thee 'Charles'," that is a holy ceremony, and even if the child is kidnapped from the baptism site and known for the rest of his life as "Harold", God will know that he is Charles. But a stipulative definition is not a baptism. The property has to answer to its "name" in various ways if it is to be true that the word means it, and it is connoted, not denoted, by the word. Still, the nominal definition, that "Boche" means a cruel and barbarous person, could be true. The required responses are possible.

But how could "Boche" then apply truly to all German nationals as a matter of meaning? That an ignorant and angry soldier might believe that it so applies does not make it true as a matter of meaning. The rule might be "'Boche' applies to all German nationals as a matter of meaning." Or the rule might just be that it applied to all Germans and then it would be added somewhere that that is a rule of meaning.

Here it would seem that Dummett differs from Putnam and Quine. For they reject the distinction between a rule of meaning and a general law. My proposal is that there is a distinction, but that it is not properly marked by taking a rule that sounds just like a general law and labeling it a "rule of meaning". The homely format of a nominal definition is one many philosophers have been taught to despise as part of the development of scientific "semantics" and "cognitive science". But it is not easy to confuse it with a general law. Sentence (2) has the tone of a general law. But sentence (1) makes it clear to us that, if (1) is true, what (2) expresses is no merely falsifiable general law.

If we are allowed to press our request that the claim about "meaning rules" for "Boche" be backed up with a good old-fashioned nominal definition, we get a questionable but meaningful one for the inference from "Boche" to barbarity, but what do we get for the alleged additional inference from being a German national to "Boche"?

Before pursuing this question, it may be worth going back to the point that there are many meanings of “meaning” (I have explained the meaning of my use of “meaning” and it is in this meaning that I say “meaning” has many meanings) and applying them to the meaning of racial and ethnic slurs. It is quite natural to speak of “rejecting” a term such as “Boche”, or “Dago”, “Spic”, “Gringo”, etc., or of deciding whether to “adopt” the term. The use of such terms implies acceptance of certain attitudes. I do not call Germans “Boches” or Italians “Dagos” and those who do express a certain hostility and contempt, at least in “primary uses”. Hopefully “primary” is not too unclear here — such terms are sometimes used in “friendly” jokes and perhaps sometimes with success to indicate a friendly attitude.

Such terms do express properties. The Mexican who says, “A couple of Gringos just came in”, is asserting that a pair with the property of being Americans have come in. But it is not *asserted* that the speaker is in the relation to the arrivals that is *implicit* in his use of the term “Gringo”. I can unproblematically refer to myself as an American, but there is a certain oddity in my referring to myself as a Gringo. (Or should I say, referring to myself as a “Gringo” — this too is a matter for later discussion.) And that is because (to engage in the use just labeled odd) I *am* a Gringo. It isn’t impossible for me to call myself a Gringo — or for a Norwegian who has never been out of Oslo, has no Latin friends, etc., to call me a Gringo — but there is a certain oddity.

“Certain oddity” has a lameness about it which I will not attempt to eliminate. It is associated with such notions as “pragmatic” or “conversational” or “contextual” implication, where it is said, for example, when I say, “He is a Gringo,” I assert that he is an American, but in using “Gringo”, merely imply that I am not an American and may have various attitudes towards him, etc. This “implication” is weaker than assertion in that there is no contradiction in my going on to say that I myself am an American, as there would be if I were to go on to say that he is not. But there is enough force in the implication to create a “certain oddity”, which might be used for some clever purpose (humor) but might also reflect a lack of understanding.

This lack of understanding might naturally be called not understanding the “meaning” of the term. And this brings us back to the topic of “meaning rules”. One might naturally say that a rule for beginning users of “Gringo” is that it applies to Americans and implies being arrogant and selfish. I am not in a position to deliberate as to whether I “reject” this term from my vocabulary, being in the target group — or rather, my rejection would involve different considerations. I might call for its elimination, especially if I went to live in Mexico. But the "meaning" involved here is not the property that is asserted of or the proposition that is asserted. It is rather a matter of things like implication.

This is vague in the extreme, and it is open to a critic to reject the attempted distinction. Here I am merely urging it for consideration, assuming that it has not been clearly recognized as an option and that its optionality can be clear without its formulation being so. The point includes the idea that "Boche", "Gringo," etc., are not fully definable as terms in general use. We should remember the warning at the beginning, that in giving the meaning, or property expressed, we must distinguish between giving the meaning expressed on some individual use and a more general claim about a property usually expressed. I do not deny that someone might in some case assert "He is truly a Gringo" and thereby assert that the referent is arrogant. We could then nominally define the term "Gringo" as used by him on that occasion as meaning, perhaps "arrogant, selfish American". But this would be an unusual case. (By "unusual" I mean not the sort that naturally occurs to me when I try to imagine examples.)

To determine whether someone is asserting that the referent is arrogant, or implying this, or implying that he has a tendency to accept a certain stereotype of the referent, or something else, may be difficult and often not worth attempting. The basis for answering such a question is testing for cancelability. Ordinarily, one can consistently say "George is a Gringo — but mind you — he's not at all arrogant" and can't consistently say "He's a Gringo — but mind you — he's not an American." But of course there is free play in language which will make the question of consistency difficult in context, and we can imagine cases where the first sentence is inconsistent and the second is not. Difficult cases for applying a distinction are not sufficient reason to give up the idea of the distinction, and it is important to understanding the notion of a definition to recognize that there are other things called "meaning" besides a property expressed. Explanations as to the typical attitudes of a speaker who uses a certain term may be called "meaning rules". Such "meaning rules" give the "meaning" in a different meaning of "meaning" than that of identifying the property expressed. The sentence "Defining is giving the meaning" can be taken in this other meaning of "meaning" to give another meaning of "defining".

There would also be new meanings for "inferable" and "language". It does not follow logically from "He is a Boche" that he is cruel and barbarous. It does follow that he is a German, and it is implied that the speaker thinks him cruel and barbarous. This is inferable "pragmatically", not logically. And the term "Boche" is part of his "language" but not of mine, because I don't use such words.

However, I understand them, and as useable in combinations that qualify them as terms of English. Here vagueness comes in again. Those pressing for the "elimination" of "gender-biased language" may wish not merely that I cease to use masculine pronouns to refer to people generally, but that I cease to understand such uses, so that they really die out. That is, when someone uses a masculine pronoun in that way, I would not merely disapprove of him but would not understand him. In this way, there could be pressure for the elimination not merely of "Boche", but of "German". For that matter, there has sometimes been a spirit pressing for the elimination of language. But there is a difference between getting rid of my capacity to assert some proposition and giving up use of terms that imply certain attitudes.

There is a certain freedom I have in choosing whether to adopt a certain style of speech (an ethnic slur is one of a great variety). There is another sort of choice I may make in constructing a formal system and adding or deleting various terms. Then there is having a term available to express a property in English. Different meanings of "language", "inferable" and "mean" are associated with these three. Running them together can make for strange and paradoxical sounding results.

Now we may return to the question of what meaning rule could bring about licensing a *logical* inference from being a German to "Boche". It may be possible to make the term "German" mean a person who is, among other things, cruel and barbarous. That is, the property of being barbarous would be part of the property expressed by "German". This would merely be a new use for "German", one already served by "German who is cruel and barbarous." "Eliminating" this latter phrase, in the sense of refraining from its use, is a matter of attitudes towards Germans. But failing to recognize even the possibility of its application with that meaning is muddled asceticism — a vow of intellectual poverty. Anyway, the possible exercise of reassigning meaning for "German" does not help with "Boche". The nominal definition, "'Boche' means a German national", effects the required licensing of logical inference, but then it either conflicts with the definition linking "Boche" to being cruel and barbarous or else merely registers another sense of that term, so that the inference from "German" to "barbarous" via "Boche" would be an equivocation.

This is not to say that two different definitions for the same term with the same meaning cannot be given. This is a point of weakness in my scheme of things. The distinction between two ways of speaking of the same property and speaking of two necessarily coextensive but nonetheless distinct properties is not always clear. "'Brother' means a male sibling" and "'Brother' means a male child of the same parents" do not represent a disagreement over the meaning of "brother" or two different senses of that term. "'Four' means twice two" and "'Four' means half of eight" do conflict as definitions, even though they are in perfect harmony in terms of the sentences they license as expressing necessary truths. Mathematical practice is not disturbed by counting logically equivalent definitions as the same, but this is not acceptable for philosophy. But these problems do not bear on the point that two nominal definitions assigning as meanings properties that are not even logically equivalent represent either different senses or a conflict about the meaning of the term defined.

Consider then: "Boche" means either a German national or a person otherwise qualifying for the term "Boche". This attempts compatibility with the definition linking "Boche" to barbarity, but at the expense of another failing. The appearance that we can say such a thing as that "Boche" means a person qualifying for the term "Boche" may be fostered by ignoring the distinction between the triviality that "Boche" applies to all to whom "Boche" applies and the falsehood that "Boche" expresses the property of being a person to whom "Boche" applies. To apply, in the relevant meaning of "apply", namely, truly or falsely, to anything, a term must express a property, and it cannot, without regress, be the property of being a thing to which the term applies.

Again: "Boche" means either a German national or a person otherwise qualifying as a Boche. This doesn't qualify as a definition either, because of its blatant circularity. This is a matter of degree, which is always a sign of trouble to a platonist. But "'Boche' expresses the property of being a Boche" would be true if only there were such a property as being a Boche (if only "The property of being a Boche" meant something — a distinction still needing discussion). Even then it would not be a definition. Anyway, there is no such property; "Boche" does not express any property beyond being a German. So the above sentence fails even to express a false claim about the meaning of "Boche".

I conclude, then, not only that there is no meaning rule, in terms of a rule giving the sense of an expression, that makes being barbarous follow from being a German, but that there could not possibly be such a rule. In making up a formal system, one may have to worry about producing a "non-conservative extension", or a formally inconsistent system. But we can contribute to the evolution of our natural language without fear of its ever having such features. To make a term express a property is somewhat in our power, somewhat not, a complex matter, but success at doing so can never make any difference to what content is inferable from what.

Some philosophers have held that the difference between real and nominal definition is merely a matter of different styles of quotation, direct or indirect. The nominal definition, that the word "bachelor" expresses the property of being an unmarried male, would on this view differ from the real definition that the property of being a bachelor is the property of being an unmarried male, merely in that the first uses indirect quotation while the second dispenses with it.

This view runs into conflict with the recognition that the real definition is necessary while the nominal definition is contingent. This would seem to show a deep logical difference between the nominal and the real definition, which is explained with great force and clarity by Casimir Lewy [Lewy (1976)]. In reply to this difficulty it might be held that to say that it is necessarily true that all bachelors are unmarried is merely a way of saying that the word "bachelor" means an unmarried man. Talk of necessary truth is a misleading way of talking about the meaning of words. That it is not necessary that the word "bachelor" means an unmarried male would then merely reflect the fact that what is necessary is not necessarily necessary. If it were to cease to be the case that the word "bachelor" means an unmarried male then it would cease to be the case that it is necessary that a bachelor is an unmarried male.

One problem for this reply would be making it into a general account of all claims about necessity. The necessity of the claim, for example, that an unmarried male is an unmarried male would seem, following the above rule, to reduce to the fact that the phrase "unmarried male" means an unmarried male. And this is absurd. It is possible that the phrase "unmarried male" should cease to mean an unmarried male, and this would, on this account, entail that it ceased to be necessary that an unmarried male is an unmarried male.

Perhaps it would be denied that this is absurd, or it might be proposed to deal differently with the necessity of "logical truths", as in Quine's discussion, where the alleged necessity of analytic truths is explained (to Quine, unsatisfactorily) in terms of reducing to logical truths via definitions, and then the necessity of logical truths, or their being "logical truths", is explained (to me, unsatisfactorily) in terms of their being instances of sentence forms (with no non-"logical" constants) that have, as a matter of fact, only true instances.

The account presently considered could look for its own way of dealing with the necessity of logical truths. It might even not account for their necessity, but focus merely on rejecting the claim that real definitions are necessary truths in the same sense that logical truths are.

Suppose that we succeed in introducing the term "alpha" to mean a bachelor, that is, express the property of being a bachelor. Then this nominal definition entails that the sentence "An alpha is a bachelor" expresses a necessarily true proposition. It does not necessarily express even a true proposition, but the proposition which as a matter of contingent fact, it expresses, is a necessarily true one. The preceding account suggests that the claim that the proposition that an alpha is a bachelor is necessary is just a way of reporting on the nominal definition, i.e., that "alpha" means a bachelor.

Against this it could be said that the proposition that an alpha is a bachelor is just the same proposition as the proposition that a bachelor is a bachelor. The nominal definition merely makes clear that "An alpha is a bachelor" and "A bachelor is a bachelor" are two ways of saying the same thing. The real definition is not, on this reply, necessary in a way different from a logical truth. In the account under discussion, the necessity of a real definition is the product of convention in that calling the real definition necessary is just a way of reporting a convention (nominal definition). In the reply, the convention merely brings about the expressing of a necessary truth, but its truth does not result from the convention any more than it does for logical truths.

Thus the idea that the meaning of a term — the property or relation it expresses — is determined by convention conflicts with the idea that necessity is determined by convention. If changes in conventions made it correct to call some married men "bachelors", this would tend to show that the necessity of the proposition expressed by "All bachelors are unmarried" is determined by convention only if those changes in conventions did not change the meaning of that sentence. If they were a change in its meaning, then the necessity of the original proposition could remain. The proposition just wouldn't be expressed by that sentence anymore. Even if it ceased to be expressible at all, that would not mean that it ceased to be or to be necessary. For convention to effect the necessity it should not effect the meaning.

Suppose that a convention succeeded in changing the meaning of "bachelor" so that it applied solely to bachelors of arts. "All bachelors are unmarried" would no longer be true. It could be held that this showed that there has been a change in the necessity of something, but this is an extremely implausible claim. Those with complexes about ontology will want to find some way of phrasing it "noncommittally", but everyone will admit it natural to say that such a convention has changed the meaning in such a way as not to reflect at all against the idea that the original sentence expressed a necessary truth which has not itself changed truth value.

On the other hand, suppose that a campaign begins for linguistic reform to allow some married men to be counted as bachelors by virtue of their life style. What one is called does matter. There might be strong feelings for or against classing some men as bachelors. Some married men might complain bitterly about their being made to feel second-class citizens in singles bars. Their remedy would be to class themselves as bachelors in spite of their having wives, holding that only linguistic reform can change the unjust perception of them as men who belong at home. While this involves the mere word "bachelor", it also involves the linguistically identified role. The debate could be international and interlinguistic. The result could turn out so as to provide a case for those who deny that necessity must always be itself necessary, if possessed at all. In the case of the conventional change to bachelors of arts, the meaning of the sentence changed. In the case presently described, the meaning would stay the same while the truth value changed. It would seem plausible to say that in this case something which was once necessary ceased to be, that it has come about that there are now some married bachelors. By "plausible" I do not mean "true". I think this is a false view, but an understandable one with some plausibility.

Against this plausible view it might be objected that if it were true that being a bachelor is a social role which could be extended to some married men while remaining the same role, then this shows not that the necessity of the proposition that all bachelors are unmarried is nominal, contingent on facts about words (in the broad sense of "linguistic practice" as opposed to mere facts about the English word "bachelor") but rather that the proposition is simply not necessary. This reply could preserve the principles of modal logic by adjusting their application. That it is after all possible for a bachelor to be married would be surprising, perhaps, but then instead of that proposition being necessarily necessary, it would be necessarily possible. We would have necessary truths and iterated necessity, but starting from a different base.

If it were true that the institution of bachelorhood could persist through a change allowing married men to qualify as bachelors, a bachelor could not be defined as an unmarried male, and "bachelor" could be defined to mean an unmarried male only by ceasing to be title word for the institution. It might then be asked how, in this hypothetical case, the word "bachelor" could be defined.

It would be easy to plead that the case is so counterfactual that one can't imagine what could be meant by "bachelor" as the title word for an institution which could change from not allowing the married to qualify to accepting them. Suppose, though, that the new qualification is being a man not having a certain relationship to a woman of sexual fidelity or a certain complex pretense of fidelity. A traditional-style clandestine philanderer would not qualify as a bachelor, but a man in an "open marriage" would. The details won't be worked out here. Let us call the condition "non-monogamy". Then it seems that we could nonetheless not define "bachelor" for this case as meaning a non-monogamous male. For the institution could, presumably, change back into what it now is. So what common property could be marked by "bachelor"?

My account of meaning is committed to the existence of properties and the relation between a term and a property — of the former expressing the latter. This does not entail commitment to there being a common property expressed by "bachelor" in the two cases discussed above. But it may seem that if the cases are connected as hypothesized, and are possible, then there should be a common property involved. For some men will become bachelors under the new scheme and others won't. That is, the new scheme, not marriage or divorce, will change the status of some but not others. We will have to say such things as "He is still a bachelor" and "He is not a bachelor anymore." So what basis could there be for resisting the translation to "He still has the property of being a bachelor"? Well, why should I worry about whether there is a basis?

Howard Wettstein cities as a "fundamental criticism of Fregeanism . . . that *There can be no Fregean sense of terms like 'game'*" [Wettstein (1988), p. 13, n. 31]. While I prefer the title "platonist" to "Fregean", it would seem the criticism, whatever it is, would be the same. But what is it?

Suppose that Hiram is at table, absorbed in a chess game with Horace. His wife Helen is sitting beside him, watching. Under the table, her right shoe is off, and she is playing footsie with Horace. Her mother-in-law, observing the trio from across the room, exclaims "Hiram, you're lucky to have a wife with the patience to sit there while you ignore her for your tiresome game." Helen says to herself, "Don't worry, Mother dear, I've got my own game going." Do she and Hiram share the common property of being absorbed in a game? Is there anything common to those two activities besides their both being called "games"? Why should I care about this?

Suppose the answer is yes, they both have the property of being involved in a game, that is, both are playing a game. Could the charge be that this is to see more in common in this case than there is? Wouldn't the remedy then be to deny that it is more than a joke to say each is playing a game? (We may recall the joke about coming home in a flood of tears and a sedan chair.) Perhaps then it would be said that this line would rapidly lead to a dozen different "senses" for "game", with the feeling that it was not clear where to draw the line between expressing a common property and expressing distinct properties. The limitless variety of uses a single term can have, it would be said, do not fit into the clumsy subdivisions required for distinctions of sense.

This strikes me as close to the anxiety of the "desert landscape" ontologist who objects simply on grounds of the number of "entities" required for the distinctions and the difficulty of drawing them. Perhaps it is difficult to determine whether two uses attribute a common property. It may call for subtlety. Is that intolerable?

This is not to say that the question whether two uses attribute a common property is always interesting or answerable, or still less, both. It may sometimes be a hopeless or idle question. But that is no reason for making the question unraisable as a matter of ontological principle, ruling that there is no such thing. Whether a certain dog is the property of Jones or of Smith may be a hopeless or idle question. That doesn't license the conclusion that there is no such thing as ownership.

The complaint about seeing more or less in common to different uses than there is should not license giving up on the question whether different uses do have something in common. This does not require that what the different uses have in common is that they express a common property. But what they have in common will nonetheless be a common property. For that is what having something in common in this kind of case (as opposed to sharing apartments, spouses, boundaries, etc.) is. And this kind of having in common, like the others, can be expressed in language.

If the good boy spends the afternoon depositing collections from his extensive newspaper route, while his lazy brother goes fishing alongside the river, have they both spent the afternoon at the bank? Is that one common feature of their different performances, in addition to the fact that the word "bank" applies to their two locations? I would say no, that the common property is one of being at places to which the word "bank" applies, but in different senses. And that is not just a matter of being a thing to which the phrase "was at a place to which the word 'bank' applies" applies. An infinite regress here is avoided by the nominalist only by some improbable ruling, say, that we stop conflating properties with words once we get to properties of words.

Our habit of ignoring the difference between being at a place called a "bank" and being at a bank is understandable, but can cause confusion in logical puzzle cases. "Was Sam at the bank?" "It depends what you mean by 'bank'." This is confusing. It's not whether Sam was at the bank that depends on what you mean by bank, but rather where you were asking whether he was at.

To say "The truth or falsity of the proposition that *P* depends on the definition of the term *t*" is usually incorrect, unless *P* is about the definition of *t*. It is a way of speaking that purports to identify a proposition while claiming that what the proposition is, is not yet determined. If the proposition had been referred to as "the one Bill was considering" we could properly go on to determine what proposition this was, but so-called "indirect quotation" is a source of deep misunderstanding. It purports to represent the content while "direct quotation" represents the words. But to know the content is often a difficult, sometimes an impossible, task of interpretation, while using the device of indirect quotation is simple and may tend to conceal this difficulty.

Thus, if we speak of "Hegel's proposition that the Absolute is the World Process", this form of speech purports to present a content when we might more properly wonder if there is one to present. To avoid indirect quotation and speak instead of words of Hegel's which translate into English as "The Absolute is the World Process" is tedious. The ideal of the form of speech, indirect quotation, is regularly ignored for the sake of convenience.

"For the sake of convenience" is oversimple. People do not ignore the ideal purpose of indirect quotation so much as they fail to be aware of it. It is difficult to be a spokesman for this ideal in an environment in which people are conditioned to reject it without ever understanding it. (Which is itself a peculiar, interesting sort of condition.) "The proposition that Bill was considering" does not purport to represent Bill's proposition. "Bill's proposition that The Absolute is The World Process" does purport to do so. It is all right to say, "I don't understand at all the proposition that Bill was considering." To say "I don't understand at all the proposition that The Absolute is the World Process" is perfectly acceptable in standard usage. But ideally it should be put in direct quotation or else in terms of "The proposition Bill claimed to be considering." Ideally, the indirect quotation represents the proposition as understood to a certain extent.

"To a certain extent" is very lame. Defining a proposition can be done better or worse. Our grasp is chancy and a matter of degree. The platonic doctrine that meanings are eternal and unchanging does not mean that our grasp of them is. At best, in a good dialectical process, there are periods of understanding. But there is nothing eternal about such periods. We need practice or we become rusty and fatigue or age will always interfere eventually. But to be realistic about our prospects for understanding does not require despairing of them.

If I define "Utilitarianism" as the name for the proposition that what is good is what most efficiently promotes the general happiness, then this should (of course it doesn't as things are) signal my understanding that proposition, whether accepting it or not. Then that proposition could be defined by representing it as a compound of some better understood propositions.

Against this it may be observed that many thinkers have called themselves by the name "Utilitarian" and have given different, even mutually incompatible, accounts of what this doctrine is. If we require a definite proposition to be the doctrine, then we may fail to understand the movement, which is not acceptance of this or that definite proposition, but rather, perhaps, a tendency to produce or accept claims that share a certain family resemblance.

This is a good thing to remember. The allegiances of men to various opposing causes are a principle factor in human history. Even in the most abstract and esoteric studies, the tendency to break into factions retains more than just a trace of the importance it had when it was a matter of life and death battles. This rallying to positions not only does not require accepting definite propositions, but may even be impeded by it. It is possible for a man to spend his life without ever grasping a definite proposition and be nonetheless prominently contentious and partisan, perhaps even achieving great worldly success.

Nonetheless, the effort to formulate and understand definite propositions is a vital part of healthy intellectual activity. This understanding should include knowing the sort of interconnections recognized in defining.

That no one *ever* understands any proposition is a false claim representing unhealthy pessimism or confusion. Unfortunately, the phenomenon of thinking one grasps a proposition when one does not is common enough to give the very notion of understanding a bad name. Do I understand the proposition that what is good is what most efficiently promotes the general happiness, as I signal that I do in using the device of indirect quotation? Can I offer that "proposition" to define the term "Utilitarianism"? And what proposition would it be, which could then be further defined? Various, incompatible "versions" of Utilitarianism have been presented. Surely these are not various definitions of some one proposition?

Of course they are not. The traditional Utilitarians were not clear in their own thinking whether they were claiming that the property of being good is one and the same as the property of most efficiently promoting the general happiness, or some other of a variety of propositions equally well suggested by the words they used. And what about the words I use, in a pretense of transcending the unclarity and showing at last the options? Real dialogue involves the advance of understanding, and we have the grammatical signals to serve as marks. But we often talk as if we understand when we do not. I do not care for the "Whorfian hypothesis", which has had less clarification than "Utilitarianism", but if any idea really is built into our language, it is optimism about the possibility of communication of content, of propositions. A pragmatically motivated nominalism faces a hopeless task, since it is such a pessimistic one — a repudiation of the features in our language which signal grasp of content.

In this connection, it is worth considering another complaint implicit in some nominalist discussion of the doctrine of meaning as sense or connotation. The doctrine has it that a term which expresses a sense or property denotes or applies truly to a thing if and only if the thing has the property the term connotes or expresses. But then it seems to some thinkers that this conflicts with the fact that it is *we* who decide whether a term applies, in our free use of our language. Definitions are also made out by these thinkers as just another attempt to restrict human freedom. The crusading philanderers wanting to be called bachelors will point out that it is a name and that we are free to apply it as we decide. We are told how a craving for security, a quest for certainty arising from neurotic anxiety, can lead to the acceptance of intellectual slavery as an escape from facing our existential freedom to apply our language as we choose. Rather than asking whether a term applies, we are supposed to ask whether we apply it. Rather than asking whether a certain thing is a dog or whether the term "dog" applies to it, we are to ask whether we call it a dog or we apply the term "dog" to it, with the requisite general agreement, pragmatic success, etc.

These thinkers work the border zones, citing borderline cases as especially tempting examples in which it is we who decide, not platonic meanings and logical laws. We then get a situation in which people are urged to repress the impulse to think that it has nothing to do with us whether there ever were, say, oak trees or brontosaurs. The cure for this impulse is contemplating borderline cases of these things, where it seems natural to say that it is for us to decide, rather than being a matter settled between the property of being an oak tree or of being a brontosaur, and the laws of logic.

This is a caricature (though a caricature of a caricature), and nominalist philosophers will themselves want to resist the suggestion that we are completely without constraint in applying our language. The constraint offered is social — a rejecting of the absurd idea that individual choices of word use are without constraint. Crusades can still be mounted. But the nominalist needn't be worried about being embarassed by any crusade for language reform, since the embarassing ones will be, at least while embarassing, unsuccessful, and thus fail the social criterion for applicability of terms.

Consider the property of being Miss America for the year 1949, or being 1A for the draft in that year. We have terms to express these properties, which obey the laws of logic. Does recognizing this entail failing to see that whether a thing has one of these properties is, at the appropriate time, up to the appropriate people, the judges or the selective service board, or perhaps to some higher authority to which an earlier decision has been appealed? Is the platonist on the wrong side here, along with evaders of responsibility, the judges or board members confronting an angry mob of decision protestors saying, "Don't look at us, we just work for the system, applying the definitional criteria"?

If we are asking whether *A* shares with *B* the common property of being a tadpole (and *B* is indisputably a tadpole) or rather, shares with *C* the common property of being a frog (and *C* is indisputably a frog), we may find a certain tendency to think of the question about *A* as one that is up to us, while the question about *B* or *C* is not. Then the nominalist could be seen as pushing for extending the attitude about *A* to the cases for *B* and *C*, and the platonist as pushing the opposite direction. Seen in this way, both sides are pushing against common sense, which goes one way for the borderline case, another for the paradigm.

Here it would be unjust to saddle the platonist with the dubious idea that the forms of words "*x* is up to *y*" or "*x* depends on *y*" express just one relation. To note that they may mean various things and in some problem cases, nothing, may help with this puzzle.

The doctrine that there are properties and that it is possible, on fortunate occasions of communication, to express them with words and to thus attribute them truly to things is compatible with the view that the property so expressed may be one possessed thanks to a decision by us. It is even compatible with the view that the same property may be one which is possessed by one thing independently of us and yet by another thanks to our decision.

This is a peculiar view which I do not endorse but is worth stating. It could be held that before languages developed, there were dinosaurs that were herbivorous, and that there were others which were borderline cases for being herbivores, such that we would have been needed on the scene to settle the matter.

That this is compatible with platonism may seem questionable. It must be stressed that this compatibility depends, first, on the borderline status of a herbivore candidate being independent of us, and, second, on its being a consequence of our not having settled the case that the candidate is not a herbivore. This might seem to contradict the idea that the case would be up to us, but it does not. A Miss America candidate is not a Miss America until the judges decide that she is. So before they decide, she is not. Her case is in that sense settled. We do not say that the judges reopen her case and revoke her position as a non-Miss America, when they vote her the title. But still, in the absence of the decision, she does not have the title. It is not that she neither has the title nor lacks it, as some "alternative logicians" would say. Or, at least, we do not need to say this to do full justice to the fact that her having the title is determined by a human decision.

Similarly, it can be said that *A* (at time *t*, of course — *A* will have died a frog long before this dispute is over) requires our decision as to whether he shares with *B* the property of being a tadpole, while *B* (as of *t*) is a case that is, as it were, out of our hands. This requires (for a platonist) that, absent our decision, *A* is not a tadpole. That is required to give content to the saying that our decision is the key factor. But *A* could be a tadpole by our decision, on this line.

This line is, to me, quite unappealing. It has been thought to solve puzzles involving borderline cases and sorites. It does not, because marking our decision as the key pending factor in a borderline case does not justify suspending the logical reasoning that requires a break-point in a sorites series [Cargile (1969)]. The logical revisions can be made arbitrarily and successfully if done with skill at the formulation of formal rules. But they do not get a rationale from the idea that our decision makes the difference, an idea compatible with a platonistic classical logic.

It is usually more congenial to me to say of a borderline case that we do not know whether the term applies than to say that we have not decided and thus determined its range of application. Either way, it will seem that the borderline case counts against the definability of the term, since the definition represents the property expressed as a compound of better understood properties and thus would seem to promise a resolution of the borderline question.

It might seem that since defining is representing as compound, we would have to accept logical atomism as a condition of definability. Leibniz argued forcefully that, if anything is compound, something must be simple. Combined with the view that the simple property must be such that whether it is possessed or not can be known directly, foundationally, we get a view which does seem inadequate to account for the perplexities over application that arise in real life. This then may be cited as a reason for rejecting the very idea of a real definition and the doctrine of expressing compound properties in terms of which it is formulated.

The better course is to mistrust the assumptions, first, that a simple property must be knowable either directly or not at all and second the alleged connection between compounds and simples. Rejecting the first assumption allows keeping the idea of a simple property. It might be said, for example, that the property of being good is simple and yet knowable sometimes, sometimes not — and often only indirectly. Rejecting the second assumption allows giving up the notion of a simple property altogether. To be a brother is a compound of being a male and being a sibling. But being a sibling is a compound of being either a brother or a sister. Thus, being a brother is part of being a sibling, and being a sibling is part of being a brother. A compound property is the result of relations between other properties such as conjunction, disjunction, conditionality, etc. The things thus related need not be simpler, in some absolute way, than the property of which they are components.

That a property is represented as compound does not even give a clear meaning to the saying "it is really compound." That a property can be represented as compound may even be said to be relative to language in some way (explaining what way is a task for those who wish to say this). Even though a real definition is usually not about language it is always presented in language and is a performance that may be possible in one language and not in another. That this latter should be so is doubtful to me. But even if some real definition can be given in one language and not in another, this is no reason for suspecting that the project of real definition is in general hopeless.

Compounds are not a simple topic. Not every presentation of a compound will count as defining. For example, any property can be presented as a compound of a long disjunction of arbitrary properties conjoined with the negations of all of them but itself. The compound does not guarantee improved understanding. Improvement of understanding is as much a criterion of good definition as it is a product of it.

The logical atomist programme not only had an unwholesome notion of an absolutely simple property, but a far worse, empiricist conception of the nature of such allegedly foundational properties. The question as to what properties we do know best and do best to know is crucial to any theory of knowledge that can elevate the soul. Only from that standpoint — which has not been approached here — could we begin to see the true value of defining.

There is a sort of example popular with those who downplay the role of definition in understanding. On the one hand, an "A" student in law school knows the definition of "manslaughter." On the other, a crafty veteran lawyer knows what is called "manslaughter" in what courts, by what judges, under what circumstances, etc. When a question arises as to whether we have here a case of manslaughter, whose advice to we prefer — the young definer or the crafty lawyer, who may (for purpose of example) even be a bit rusty on the textbook definition? It is implausible to suppose the crafty veteran to be ignorant of the definition and implausible to suppose that understanding it has nothing to do with his qualifying as an expert on manslaughter. But it is not so implausible that the young student understands the definition as well as the veteran and yet does not understand manslaughter nearly as well.

However, there are alternatives to this plausible description. We might say that the student doesn't understand the definition as well as the veteran, challenging the idea that understanding for definitions is relatively quicker and easier (at least for the clever) than the understanding won through long experience. Or we might say that what the crafty lawyer understands so well is what will get called manslaughter among his network of acquaintances (which happens to coincide with those who have power in this matter), and not manslaughter itself. But this might encourage the reply "so much the worse for the notion of 'manslaughter *itself*'." It would be understandable in this case to think that what the people in power call manslaughter is what is important.

One might be led in this way to a version of "nominalism", according to which the important question as to what properties a thing has is reducible to what it is called. Being called a murderer by the right people is what sends a man to prison far more surely than murdering.

For all that, most of us will draw the distinction between being guilty of manslaughter and being found guilty by a court. It is harder, though, to go further and distinguish between being able to know that a certain individual is guilty of manslaughter, and to know what it is to be guilty of manslaughter. It is further still to see that a firmer grasp of the latter sort of knowledge may make us rather less than more confident of attaining the former sort. We may become obsessed with just beating the rap, or with getting a conviction, depending on the accident of our allegiance, and develop a "tense and bitter shrewdness," rather than cultivating that attidude towards truth which is "the antithesis of a slave's."

I am indebted to Richard Rorty and to J. P. Downey for helpful discussions of this paper.

Department of Philosophy
University of Virginia

REFERENCES

Cargile, J (1969), 'The Sorites Paradox', *British Journal for the Philosophy of Science* **20**, 192–202.

Dummett, M. (1977), *On Frege: Philosophy of Language* (New York: Harper & Row), p. 397.

Lewy, C. (1976), *Meaning and Modality* (Cambridge: Cambridge University Press).

Putnam, H. (1966), 'The Analytic and the Synthetic'. in *Minnesota Studies in the Philosophy of Science,* Vol. III, H. Feigl and G. Maxwell, eds. (Minneapolis: University of Minnesota Press), pp. 358–397.

Soames, Scott (1984), 'What Is a Theory of Truth?' *Journal of Philosophy* **81**, 411–429.

Wettstein, H. (1988), 'Cognitive Significance Without Cognitive Content', *Mind* **97**, 1–28.

JAMES H. FETZER

PRIMITIVE CONCEPTS: HABITS, CONVENTIONS, AND LAWS

Perhaps no aspect of the theory of definitions has become more familiar to students of this subject than that there are really only two ways in which every word that occurs within a language could be defined. The first — that of definitional circularity — arises when the words that occur in a language *L* are permitted to be defined by means of other words, which are ultimately defined by means of those original words themselves. The second — that of definitional regress — arises when new words are allowed to be introduced to define the meaning of old words, and new words for those, *ad infinitum*.

On first consideration, the prospect of languages affected by circular or by regressive definitions may look like a relatively minor inconvenience of slight consequence to understanding the nature of language. The dictionary for an ordinary language, such as English — *Webster's New World Dictionary* (1988), for example — appears to succeed in providing useful definitions for the terms that it contains in spite of resorting to definitional circularity. If there is a problem here, therefore, then it needs to be made apparent, because there seem to be no problems in practice with dictionary definitions.

The problems arising here might or might not be practical, but they are important problems, especially with respect to understanding the nature of definitions and definability. Ask yourself, for example, how you could possibly know the meaning of a defined term without knowing the meaning of those terms by means of which it is defined. Presumably, you might know the meaning of a word on the basis of such a definition or you might know its meaning apart from any such definition. Yet it is not obvious how you could possibly know the meaning of a word without knowing its definition.

Even though it may not be obvious, that it must be possible to know the meaning of some words without knowing their definitions is not difficult to establish. It might be the case, after all, that one meaningless sequence of marks was "defined" by means of another meaningless sequence of marks without rendering either sequence meaningful.

J.H. Fetzer, D. Shatz and G. Schlesinger (eds.), *Definitions and Definability: Philosophical Perspectives*, 51–68.

Suppose "ab" were defined by means of "cd", "cd" by means of "ef", *ad infinitum*. In a case of this kind, these "words" would have meaning only if at least one member of that sequence had meaning. These could be defined, yet still meaningless, words.

If definitions are ever successful, therefore, it must be because at least some words can be understood without the mediation of other words. The success of definition, in other words, appears to depend crucially upon the possibility of understanding the meaning of words that are not defined by means of other words. So, if the vocabulary of a language *L* consists of its defined words and its undefined (or "primitive") words, then it appears as though understanding any of its defined words depends upon understanding its undefined words. But how is it possible to understand primitives?

This essay has the purpose of exploring several different ways in which it might be possible for words to have meaning for the users of a language, even when they are undefined. There seem to be at least three hypotheses:

(*h1*) their meaning is determined by natural laws;
(*h2*) their meaning is determined by personal habits; or,
(*h3*) their meaning is determined by social conventions.

The problem, however, cannot be solved merely by accounting for the meaning of words. Different words can have the same meaning even though they were never so defined. The solution requires appeals to primitive concepts.

Sections 1 and 2 focus on the relationship between words and meaning where it becomes increasingly evident that an adequate theory of the meaning of primitive words depends upon an analysis of personal habits, tendencies, and dispositions. Section 3 explores several examples. Sections 4 and 5, however, shift attention to the relationship between meaning and reference, where the role of social convention in promoting communication and cooperation within communities becomes central. Sections 6 and 7 finally consider some relationships between social conventions and natural laws.

While the fundamental unit of meaning for a language user *z* appears to be a function of that user's linguistic habits, tendencies, or dispositions (see Fetzer 1989 and 1991), the fundamental unit of meaning for a community of language users *Z* is a function of those users' shared linguistic habits, tendencies, or dispositions. The success of efforts to ensure that different members of any such community share the same linguistic habits, tendencies, or dispositions, however, appears to depend upon the existence of natural laws of various different kinds, which are presupposed by linguistic conventions.

1. MEANING OF PRIMITIVES AS DETERMINED BY NATURAL LAWS

The most conspicuous advocate of hypothesis (*h1*) continues to be Jerry Fodor, who has advanced a theory now known as "the language of thought" in several publications (Fodor 1975 and 1987, for example). According to this conception, every (neurologically normal) human brain is born with an innate set of psychological primitives. In order to learn a natural language such as English, therefore, a person *z* must discover how to "match up" the words that occur within that natural language and the concepts that occur within the language of thought. The only problem is to pair them up right.

Fodor acknowledges that more than one word in a natural language can have the same meaning as a concept in the language of thought. Moreover, this is indispensable to the adequacy of his theory, since otherwise it would imply that only *one* natural language could be successfully "paired up" with the language of thought. If that were the case, then every (neurologically normal) human being would have to speak the same natural language, as a matter of natural law. Fodor's theory is thus compatible with the existence of a variety of different natural languages, which otherwise would refute it.

Since this approach affords a solution to the problem of accounting for the meaning of the primitive words that occur within a language, it clearly has something to recommend it. If every (neurologically normal) human brain *were* born with an innate set of psychological primitives, that *would* supply an explanation of how it is possible to learn a natural language. Indeed, Fodor maintains that it is only possible to learn a natural language on the basis of an unlearned innate language, "the language of thought" itself.

The key to Fodor's position, therefore, is the assumption that learning a language involves learning the truth conditions for the sentences that can occur in that language: ". . . learning (a language) *L* involves learning that *Px is true if and only if x is G* for all substitution instances. But notice that learning that could be learning *P* (learning what *P* means) only for an organism that already understands *G*"

(Fodor 1975, p. 80). Fodor takes for granted that the kind of understanding that such an organism must possess has to be *linguistic*. It is for this reason that what is innate must be a language.

What Fodor appears to overlook is the possibility that a person *z* might possess *non-linguistic* understanding of the *G*-phenomenon which *P* turns out to represent in *L*. For, if that were the case, then *z* might perfectly well possess one kind of understanding that is sufficient for learning *P* (learning what *P* means) without also having to possess some other kind. Those who have raised children, for example, have often noticed how typically they display patterns of behavior (playing with blocks, singing a song, drawing with crayons) before they acquire the ability to name or describe them.

Since Fodor concedes that experience and ingenuity are required in the process of matching up primitives in the language of thought with those in a natural language, the possibility remains that the "experience and ingenuity" that are required might afford a suitable foundation for a child to acquire a natural language — even in the absence of a language of thought! Indeed, to the extent to which economy, simplicity, and elegance provide criteria for preferring one theory to another in relation to a common domain, a theory that dispenses with the language of thought ought to be preferred.

The criteria of economy, simplicity, and elegance, of course, afford a relevant standard of relative preferability only when competing theories provide similar degrees of systematic power with respect to the phenomena of interest. This argument, therefore, depends upon establishing that such an alternative to Fodor's theory provides as much or more explanatory power with respect to the acquisition of language. The theory that the meaning of primitive concepts depends, not upon laws of nature, but upon habits that might vary from person to person, appears to provide a preferable theory.

From this point of view, it should be observed that, while Fodor's theory does afford a solution to the problem of accounting for the meaning of the primitive words that occur within a natural language, it does not provide a solution to the problem of accounting for the meaning of the primitive concepts that occur within the language of thought itself. Thus, a theory that accounted for both the meaning of primitive concepts and for the meaning of primitive words would thereby have demonstrated its superiority over Fodor's conception. An account of this kind is supplied by hypothesis (*h2*).

2. MEANING OF PRIMITIVES AS DETERMINED BY PERSONAL HABITS I

An adequate analysis of the nature of language ought to be approached from the perspective of the theory of signs (or "semiotic") advanced by C. S. Peirce, where natural languages qualify as exhibiting one of three basic kinds of signs (icons, indices, and symbols). Moreover, the theory of signs provides a foundation for a theory of mind, according to which "minds" are semiotic (or "sign-using") systems, some with the ability to use icons, some with the ability to use indices, and some with the ability to use symbols (cf. Fetzer 1988 and 1990). For now these important matters will be ignored.

The fundamental conception that ties these ideas together is that of the content of a concept, which Peirce accounted for in terms of its causal role:

> . . . the most perfect account of a concept that words can convey will be a description of the habit which that concept is calculated to produce. But how otherwise can a habit be described than by a description of the kind of action to which it gives rise, with a specification of the conditions and of the motive? (Peirce 1906, p. 286).

Thus, from this point of view, the theory of meaning presupposes the theory of action, where meanings are characterized in terms of their causal function in affecting the behavior of semiotic systems of various distinctive varieties.

Let us assume that human beings are influenced by their motives, their beliefs, their ethics, their abilities, their capabilities, and their opportunities (where "opportunities" concern how things are as opposed to how we think things are, i.e., they are related to various features of our beliefs, including their truth and their completeness in relation to the activities in which we happen to be engaged). People thus qualify as motive-belief-ethics-ability-capability-and-opportunity types of systems in comparison with other kinds of systems, including digital machines (cf. Fetzer 1989, 1990, and 1991).

People are *conscious* with respect to the use of symbols when they have both the ability to use symbols of that kind and also are not inhibited from the exercise of that ability, so that the presence of those symbols within a suitable causal proximity would lead — invariably or probabilistically — to an occurrence of cognition. Since an instance of *cognition* arises as an effect of causal interaction between the awareness of that sign and a person's other internal states (insofar as they make a difference to its significance for that person), the meaning of a sign cannot be fixed independent of those factors.

When a human being encounters specific words (sentences, etc.), the influence of those words (sentences, etc.) on that person's behavior depends upon their motives, beliefs, etc., specific values of which define a "context". Different persons are therefore in the same context (with respect to the use of specific symbols on specific occasions) only when their other motives, beliefs, ethics, abilities, and capabilities are the same. Then the meaning of a symbol is the same for different persons when their tendencies to behave in the same ways in its presence are the same across every similar context.

The appropriate standard of similarity of meaning, therefore, is not that different persons actually exhibit the same behavior in the various contexts in which they find themselves, but rather that the strength of their tendencies to display the same behavior would be the same were they or had they been in similar contexts. The relevant measure of similarity of meaning for specific words is not actual behavior but dispositions toward behavior. Even when different persons never find themselves in similar contexts and, as a result, never do display the same behavior, their meaning can be the same.

This account diverges from Fodor's conception of a language of thought, insofar as every (neurologically normal) human brain is born, not with an innate set of psychological primitives, but with predispositions toward the acquisition of semiotic abilities. States of mind are assumed to be properties of states of brains, where those brain states are "meaningful" because of the influence which they exert on human behavior. Some brain states may be meaningful for every (neurologically normal) human brain, while others acquire their meaning under the influence of environmental factors.

Since there are semiotic abilities of various (species-specific) different kinds, this approach affords a framework for understanding the nature of mentality for other species as well as for inanimate machines (if such a thing is possible), without presuming that minds must always be human. Some species are predisposed toward the use of icons, others toward the use of indices, and others toward the use of symbols, where these abilities can be variously acquired on the basis of type/token recognition, classical conditioning, operant conditioning, and so forth (Fetzer 1990 and 1991).

3. MEANING OF PRIMITIVES AS DETERMINED BY PERSONAL HABITS II

To isolate the content of an instance of cognition that induces the state of belief *B1*, for example, a specific context would hold constant some set of values for other beliefs *Bm, Bn*, . . . , motives *M1, M2*, . . . , ethics *E1, E2*, . . . , abilities *A1, A2*, . . . , capabilities *C1, C2*, . . . , and opportunities *O1, O2*, . . . , whose presence or absence makes a difference to the (internal or external) behavior that people tend to display in the presence of that belief. The meaning of that specific belief *B1* would then be the possibly infinite totality of tendencies for behavior across every context. (Cf. the Appendix.)

Consider, for example, the following case involving the use of language, which exemplifies the use of the same words with the same meaning, yet where the response behavior differs for two persons in different contexts:

EXAMPLE (A): A middle-aged man is crossing the street when someone cries out, "Watch out!" He looks around and sees a truck coming up the wrong side of the street, so he moves out of the way in order to avoid injury.

EXAMPLE (B): A middle-aged man is crossing the street when someone cries out, "Watch out!" He looks around but his vision is poor and he does not see the truck coming up the wrong side of the street. So he is injured.

Presumably the person described by (B) would have displayed the same behavior as the person described by (A), had he not been incapacitated.

Consider, for example, another case that involves the use of language, which exemplifies the use of different words that have similar meaning, where the response behavior is the same for people in the same context:

EXAMPLE (C): Bill has been working for his father for $5.00 per hour. When his dad says, "I'll pay you half as much to wash the car", he starts looking for the bucket and a sponge.

EXAMPLE (D): Bill has been working for his father for $5.00 per hour. When his dad says, "I'll pay you $2.50 per hour to wash the car", he starts looking for the bucket and a sponge.

Once again, presumably, anyone in Bill's context would have been willing to work at that rate of pay, where how it was described would not matter.

These are easy cases, of course, yet they illustrate the principles that are involved. Similar uses of language with similar meanings tend to produce similar behavior in similar contexts. A far more difficult case, which was posed by W. V. O. Quine (1960), is also one that ought to be familiar:

EXAMPLE (E): A native sees a white patch of fur hop by in the bushes and shouts, "Gavagai!" He repeats this each time he sees a white patch of fur hop by, and he chases after them.

A foreign translator observing this scene might find it difficult to discern exactly what the native meant, where Quine suggests three possibilities:

(a) "gavagai" means a whole enduring rabbit;
(b) "gavagai" means undetached rabbit-parts; or,
(c) "gavagai" means a stage in the history of a rabbit.

The difficulty in sorting out which of these the native might have meant has been widely supposed to support "the indeterminancy of translation".

From the perspective of the theory elaborated here, several questions must be raised. The first is whether or not these three different possible translations are significantly different or not. Different words, we already know, can be used to convey similar meanings, as (C) and (D) make clear. If these are merely three different ways of saying the same thing, then they do not represent three alternative meanings but rather three ways to convey a single meaning. If there is a difference between them, it can be made explicit by means of dispositional differences in similar contexts.

And, indeed, there would appear to be differences of this kind within the English language, where different responses might be given to similar questions as a function of differences in meaning. If there is a particular duration to a "stage" in the history of a rabbit, for example, then it would be possible to count the number of those stages that transpire during an interval of time. So there might be several "stages" where there is only one enduring rabbit. But if there were no dispositional difference, then there would be no meaning difference. They would mean the same thing.

That these meanings might differ in English, however, does not imply that they must have counterparts in every other language. Quine seems to beg the question in taking for granted that distinctions that might be meaningful in one language provide a relevant background for interpreting another. Moreover, either there are dispositional differences across different contexts or there are not. If there are, then presumably they should emerge in relation to a native language as readily as they would in relation to English. But there might be no such distinctions here at all.

4. MEANING OF PRIMITIVES AS DETERMINED BY SOCIAL CONVENTIONS

The conception of dispositions that is thereby presupposed differs from those advanced by Gilbert Ryle (1949), by B. F. Skinner (1953), and by W. V. O. Quine (1960), because those accounts were excessively behavioristic, extensionalistic, and reductionistic. They therefore fell prey to various difficulties, including some highly plausible objections advanced by Fodor and by Donald Davidson, which do not confound the present account (cf. Fetzer 1989, pp. 32–35). Even more important within the current context is that nothing about this approach precludes the possibility of private languages.

That this is the case becomes apparent from various perspectives. The meaning of the primitive words that make up a language for a specific user depends upon their causal role in affecting that user's behavior. The same words would possess the same meaning for different language users, therefore, only when they stand for the same concept and thereby exercise the same causal influence within the same contexts. But what can guarantee that the strength of those dispositions that they happen to possess in the presence of those undefined words must be the same from one to another?

Even if every (neurologically normal) human brain possesses the same predispositions to acquire the same language under the same environmental circumstances (including, for example, having English-speaking parents, attending an English grammar school, etc.), that surely affords no grounds for assuming that different children have actually been raised in ways sufficiently similar to establish that sameness of meaning. The interests of a community, however, may certainly be advanced by taking steps that will insure that different members of that community speak the same language.

Indeed, from an evolutionary perspective, it should be obvious that human communities stand to benefit when its members speak the same language. The purpose of a human community, from a biological point of view, is to promote the survival and reproduction of its members. The prospects for the survival and reproduction of its members, however, tends to be advanced by cooperation between its members. And the prospects for cooperation between those members tends to be advanced when they speak the same language, insofar as successful communication promotes cooperation.

From this vantage point, one of the most basic words in the English language is "help". When uttered with force, as in the exclamation, "Help!", it can be used to cry out for assistance. Were a word of this kind not a part of the linguistic fluency of the members of such a community, they would be incapacitated from summoning aid or from providing help when it was needed. Analogously, other basic words in the English language appear to include "fire", "food", and (notably) "yes" and "no". Even primitive human communities depend upon a shared language for their continued survival.

It should therefore be evident that social conventions fulfill a different role with respect to language than do personal habits. The purpose of social conventions appears to be to insure a suitable degree of uniformity of linguistic usage among the members of a community, where the usage of its individual members is determined by their personal habits. Thus, the evidence at hand indicates that social conventions have a secondary role to play with respect to the nature of meaning, but that they may still have a primary role to play in relation to promoting the uniformity of meaning.

If these reflections are well-founded, then several conclusions suggest themselves. The particular, in relation to the hypotheses that we are exploring,

(*h1*) the meaning of primitives is determined by natural laws;
(*h2*) the meaning of primitives is determined by personal habits; or,
(*h3*) the meaning of primitives is determined by social conventions;

it now appears to be the case that (*h1*) is false but (*h2*) is true. Since (*h2*) and (*h3*) are intended to be mutually exclusive, the apparent truth of (*h2*) entails the evident falsity of (*h3*) for understanding the nature of meaning.

Nevertheless, a certain ambiguity appears to affect these alternatives. For, even though it may be true that the meaning of primitive concepts is determined by personal habits (because meanings are fixed by means of causal influence upon behavior), it may also be true that which particular words are absorbed into those personal habits tends to be determined by social conventions. The customs, traditions, and practices that characterize specific communities, after all, profoundly affect the habits that are acquired by the members of those communities and thereby influence meanings.

5. RELATIONS BETWEEN SOCIAL CONVENTIONS AND NATURAL LAWS

Indeed, not only are social conventions profoundly important as determinants of the personal habits that are acquired by the members of a certain community, but the habits that are adopted and the conventions that prevail are profoundly affected by natural laws. The natural laws which make a difference here, moreover, appear to be of at least two distinctive kinds, namely: those that are involved in the acquisition of habits under the influence of environmental and genetic factors, on the one hand, and those that are presupposed by the existence of conventions, on the other.

That laws relating the acquisition of specific habits to specific genetic and environmental factors should be required to understand the nature of conventions is hardly a surprise. These are laws concerning the kinds of ways in which human behavior can be shaped, including classical conditioning, operant conditioning, and the like. They relate (neurologically normal) human brains of physically normal children with the acquisition of specific abilities, such as the ability to use English, under the influence of specific environmental factors, such as having English-speaking parents.

What other kinds of laws, after all, could possibly make a difference to the acquisition of conventions? The answer to this question concerns every natural law whose operation is presupposed by the establishment of a convention. Consider, for example, the distinctive aroma of garlic. Almost nothing else smells exactly like it, which makes it exceedingly difficult to describe. The expression, “the aroma of garlic”, therefore, seems to function as a primitive expression in English, where the meaning of that expression — the concept of the aroma of garlic — is its causal role in affecting behavior.

Several different kinds of laws are involved in even this uncomplicated illustration. One is that this distinctive aroma is characteristically emitted by things of a specific kind, namely: instances of garlic, which is a bulbous herb (*Allium sativum*) of the lily family. Thus, laws of the first kind relate the possession of a certain specific aroma (the aroma of garlic) to instances of things of a specific kind (garlic). Laws of a second kind relate the presence of that specific aroma (the aroma of garlic) to the occurrence of a certain smell (the smell of garlic) for human beings under suitable conditions.

The aroma of garlic qualifies as a permanent property of garlic, in the sense that nothing could lose that aroma and remain an instance of garlic, even though the possession of this property is not one of those properties that things of this kind have by definition (such as being a bulbous herb). The smell of garlic, by comparison, arises from an interaction between a thing that exudes an aroma and a thing that has the ability to detect that aroma and is not incapacitated from exercising that ability. That it has a distinctive aroma does not require that garlic ever happens to be smelled.

The use of signs of the three kinds identified by Peirce can also be exlaborated on the basis of this example. Icons are signs that stand for that for which they stand by virtue of relations of resemblance between those signs and that for which they stand. One instance of the aroma of garlic (of the size of garlic, of the shape of garlic, etc.) therefore stands for any other, although occasionally an instance of that aroma might be so pungent (or of that size so gargantuan, or of that shape so peculiar, etc.) that they qualify as “abnormal” (and thus become the subject of humor, etc.).

Indices, by contrast, are signs that stand for other things by virtue of being either causes of or effects of those other things for which they stand. The presence of garlic thus stands for the presence of the aroma of garlic, and the aroma of garlic stands for the presence of garlic. Since there can be more than one cause of certain effects (such as the taste of garlic or impressions of garlic left in the wet sand, etc.) and some effects are probabilistic rather than deterministic, some stand-for relations are uncertain rather than certain signs of the presence of those other things.

Symbols, finally, are signs that stand for that for which they stand, not because of any natural relation of resemblance or of cause-and-effect, but merely on the basis of an habitual association. In the present instance, of course, the use of the words "garlic" and "aroma" are examples of signs we habitually associate with things of those kinds. Because we happen to use these specific words as standard usage within our language community, it turns out that these words are habitually associated with these things not only by us but by other users of the same language as a social convention.

6. RELATIONS BETWEEN KINDS OF SIGNS AND TYPES OF MINDS

The use of signs always involves a triadic relation between a sign, that for which it stands, and a sign user. Thus, although icons and indices are "natural" signs in the sense that the relations by virtue of which they can stand for other things are there in nature, whether we notice them or not, we are able to use those signs only if we have the ability to detect their presence and are not incapacitated from exercising that ability. With respect to symbols, moreover, we have to acquire the associations that we create between those signs and what they stand for by means of habits.

Without the existence of something that has the capacity to use signs of that kind and is not incapacitated from exercising that ability, therefore, nothing would ever stand for anything at all for anyone (or thing). Since those things that have the ability to use signs are minds (or semiotic systems), it appears appropriate to distinguish different minds of different types on the basis of the kinds of signs they can use. Minds of Type I, for example, can utilize icons. Minds of Type II can utilize indices. Minds of Type III can utilize symbols (Fetzer 1988 and 1990).

The presence of signs of a certain kind will bring about an instance of cognition in minds with the ability to utilize signs of that kind that are not incapacitated from exercising that ability. The smell of garlic can cause a human mind to infer the presence of garlic, and the sight of garlic can be used to infer the presence of the aroma of garlic. Such inferences are by no means infallible, however, since the smell of garlic, for example, may have been produced by some unusual combination of chemicals that brings about a similar aroma rather than by the real thing.

Indeed, were there no laws that relate one property (such as garlic) to another (such as the aroma of garlic), on the one hand, or that relate the acquisition of habits of one kind (such as the ability to use English) to genetic and environmental factors (including having parents who speak English), on the other, it would be difficult, if not impossible, for conventions to be sustained. Natural laws provide universal, inviolable, and unchangeable relations between phenomena of various kinds upon which social conventions can be established. These laws are essential to their existence.

The nasal cavity behind the human nose, for example, is lined with mucosa containing receptor cells which contain microscopic hairs known as *cilia*. When they are stimulated in different ways, these receptor cells send nerve impulses to the olfactory centers of the human brain (Ackerman 1990). The patterns of neural activation that occur cause different sensations to be experienced in the human body, as an effect of the kind of stimulation that has occurred. Depending on matters of context, that pattern of activation then tends to affect the behavior a person displays.

The specific kinds of stimulation that these receptor cells experience, moreover, are brought about by the causal influence of distinct aromas. Different aromas tend to produce different smells as sensations that are experienced in a human being, when such a person has the ability to detect the presence of that aroma and he is not incapacitated from the exercise of that ability. Someone who is suffering from a bad cold complete with sinus congestion, for example, would be unable to experience the smell of garlic, even were he entirely enveloped by the aroma of garlic.

If there were no laws relating the presence of garlic to the aroma of garlic, or relating the aroma of garlic to the smell of garlic, or relating the smell of garlic to human behavior (in relation to various contexts), the word "garlic" and the phrase "the aroma of garlic" would have no constancy of meaning. The presence of garlic would be attended by various different aromas on different occasions, which, in turn, might bring about some assortment of smells that varied from one to another time. The aroma of garlic might produce the smell of cinnamon instead.

From this perspective, therefore, it is fascinating to observe that, although meaning is a function of personal habits and sameness of meaning is promoted by social conventions, our success in sustaining conventions provides indirect — but nevertheless very strong — evidence for our belief in the existence of an external world. The best explanation available for our capacity to nurture and maintain customs, traditions, and practices within a community through substantial intervals of time may even be the hypothesis of an external world governed by natural laws.

7. RELATIONS BETWEEN SOCIAL CONVENTIONS AND EPISTEMIC OBJECTIVITY

Because the meaning of primitives is a function of their causal role in affecting behavior, it is tempting to conclude that Wittgenstein was right in maintaining that we should look for the use of words rather than ask for their meaning. The uses to which words are put, however, obviously depend upon contexts. Strictly speaking, therefore, it is crucial to distinguish between *public usage* and *private meaning*, where the public use of words provides evidence for, but does not thereby define, the private meaning that their users attach to them. The relationship is more subtle.

As J. L. Austin (1962) has emphasized, people use language for many different purposes: to amuse, to deceive, to mislead, to entertain, to seduce, to impress, to confound, and so forth. His distinction between the specific words that are used on a specific occasion (as a locutionary act), the specific meaning that was thereby conveyed (as an illocutionary act), and the specified effect that was thereby attained (as a perlocutionary act), warrants contemplation. The words used on a specific occasion might be a matter of choice, but the meaning that they convey has a public aspect.

When a young woman suggested to a young man that he should not do certain things, she may have used certain words to issue a warning ("We are going too far, Harold!") and thereby have preserved her virtue (since he did not want to offend her), but still feel a little disappointed (because she had hoped that he might be more insistent). Had he perceived her real intention (her desire to preserve the appearance of virtue without foregoing the pleasures of vice), he might have known what to do (without disappointing her). Her public usage thus concealed her true feelings.

From the point of view of the theory of knowledge, therefore, what someone means by the words they use on a specific occasion could be very difficult, if not impossible, to discover, in the absence of any real ability to "read one's mind". Hypotheses about the causal role that is fulfilled by specific words for specific human beings exhibit the epistemic interdependence of motive-belief-ethics-ability-capability-and-opportunity ascriptions, in the sense that testing a specific hypothesis about one of their values requires assigning values to all (Fetzer 1989).

Without doubt, those cases in which we have reason to believe that we can succeed in understanding what others mean by what they say are those in which we know those others well. Cases of this kind ought to include parents and children, husbands and wives, doctors and their patients, for example, provided, of course, that those parents, spouses, and doctors are perceptive observers of other persons. The surprises life can bring to parties who think they know other parties well still include affairs, divorces, and suicides that were entirely unforseen.

Nevertheless, the analysis of meaning that has been indicated here supports not only our belief in the existence of an external world but also the existence of an objective conception of truth. The word "truth" no doubt, needs to be applied to our beliefs, even when we are unable to explain their meaning by means of other words, precisely because they are couched in primitive words. For if no conception of truth were forthcoming from this approach, that would be a serious shortcoming.

Once again, the argument involves an inference to the best explanation. For when we happen to succeed in attaining our objectives, the best explanation for our success may be that the world possesses some of the properties that our beliefs ascribe to it, where those beliefs are, in specific respects, appropriate to guide our actions. From this point of view, therefore, it seems plausible to assume that beliefs are true when they are appropriate to guide our actions and that "truth" itself can perhaps be best understood as appropriate belief (Fetzer 1990).

During the course of this inquiry, distinctions have been drawn between laws, conventions, and habits as possible determinants of meaning. The theory elaborated here defends the view that habits are the fundamental units of meaning for language users z while conventions, understood as shared habits, are the fundamental units of meaning for a community of language users Z. The existence of natural laws, moreover, appears to play a crucial role in ensuring that different members of a community can share the same habits, tendencies, and dispositions.

Department of Philosophy
University of Minnesota, Duluth

APPENDIX

A more formal construction employing subjunctive ($\ldots \Rightarrow$ ____) and causal ($\ldots = n \Rightarrow$ ____) conditionals for a specific mental state (such as the belief *B1*) in relation to various contexts **C1**, **C2**, ..., therefore – where n may assume the deterministic value u or any probabilistic value p – would require the use of a possibly infinite set of conditionals (S) such as follows:

(S) $$[B1zt \Rightarrow (\mathbf{C1}zt = n1 \Rightarrow R1zt^*)] \ \& \ [B1zt \Rightarrow (\mathbf{C2}zt = n2 \Rightarrow R2zt^*)] \ \& \ \ldots,$$

which would be read, "if z were in state *B1* at time t, then being in state **C1** at t would bring about (with strength $n1$) response behavior of kind *R1* at time t^*; and if z were in state *B1* at time t, then being in state **C2** at t would bring about (with strength $n2$) response behavior of kind *R2* at time t^*; and so forth – where time t^* occurs at some specific interval after time t.

Alternatively, the difference between having belief *B1* rather than *B2*, for z at t can be elaborated by holding each such context constant and by varying those beliefs in order to bring out their different consequences:

(S*) $$[\mathbf{C1}zt \Rightarrow (B1zt = n1 \Rightarrow R1zt^*)] \ \& \ [\mathbf{C1}zt \Rightarrow (B2zt = n2 \Rightarrow R2zt^*)] \ \& \ \ldots,$$

which would be read, "if z were in state **C1** at time t, then being in state *B1* at t would bring about (with strength $n1$) response behavior of kind *R1* at time t^*; and if z were in state **C1** at time t, then being in state *B2* at t would bring about (with strength $n2$) response behavior of kind *R2* at time t^*; and so forth – where time t^* occurs at some specific interval after time t.

Each of these conjuncts is s scientific conditional, where scientific conditionals are instantiations of lawlike sentences, which on this approach are trivially derivable from them by universal generalization, regardless of whether "z" is interpreted as an ambituous name or as in individual constant. The first of the conjuncts that collectively constitute the set 9S),

(SC) $$[B1zt \Rightarrow (\mathbf{C1}zt = n1 \Rightarrow R1zt^*)],$$

thus entails the corresponding (universally quantified) lawlike sentence,

(LS) $$(z)(t)[B1zt \Rightarrow (\mathbf{C1}zt = n1 \Rightarrow R1zt^*)],$$

which, of course, asserts that, "for all z and all t, if z were in state *B1* at time t, then being in state **C1** at t would bring about (with strength $n1$) response behavior of kind *R1* at time t^*". (See Fetzer 1981, pp. 53–54).

Insofar as similar formulations (involving subjunctive and causal conditionals) are applicable for the formalization of lawlike sentences generally, perhaps it ought to be emphasized that the specifically mentalistic aspects of beliefs, for example, arises from their semiotic significance for a system at a time. In other words, while there are many kinds of lawful and causal relations, only some of those lawful and causal relations properly qualify as *semiotic*. The difference between "causal systems" and "semiotic systems", therefore, arises because semiotic systems are causal systems of certain special kinds. Those who desire to pursue this issue may want to consider Fetzer (1988, 1989, 1990, and 1991).

REFERENCES

Ackerman, D. (1990), *A Natural History of the Senses* (New York, NY: Random House).

Austin, J. L. (1962), *How to Do Things with Words* (Oxford, UK: Oxford University Press).

Fetzer, J. H. (1981), *Scientific Knowledge* (Dordrecht, The Netherlands: D. Reidel Publishing Company).

Fetzer, J. H. (1988), 'Signs and Minds: An Introduction to the Theory of Semiotic Systems' in J. H. Fetzer, ed., *Aspects of Artificial Intelligence* (Dordrecht, The Netherlands: Kluwer Academic Publishers), pp. 133—161.

Fetzer, J. H. (1989), 'Language and Mentality: Computational, Representational, and Dispositional Conceptions', *Behaviorism* **17**, pp. 21—39.

Fetzer, J. H. (1990), *Artificial Intelligence: Its Scope and Limits* (Dordrecht, The Netherlands: Kluwer Academic Publishers).

Fetzer, J. H. (1991), *Philosophy and Cognitive Science* (New York, NY: Paragon House Publishers).

Fodor, J. (1975), *The Language of Thought* (Cambridge, MA: MIT Press).

Fodor, J. (1987), *Psychosemantics* (Cambridge, MA: MIT Press).

Peirce, C. S. (1906), 'Pragmatism in Retrospect: A Last Formulation', in J. Buchler, ed., *Philosophical Writings of Peirce* (New York, NY: Dover Publications, 1955), pp. 269—289.

Ryle, G. (1949), *The Concept of Mind* (London, UK: Hutchinson Publishers).

Quine, W. V. O. (1960), *Word and Object* (Cambridge, MA: The MIT Press).

Skinner, B. F. (1953), *Science and Human Behavior* (New York, NY: The Macmillan Company).

Webster's New World Dictionary (1988), 3rd College Edition (New York, NY: Simon & Schuster).

PART II

DEFINITIONAL DESIDERATA

ROY SORENSEN

VAGUENESS AND THE DESIDERATA FOR DEFINITION

The purpose of this paper is to show how vagueness is relevant to definition and thereby to thought experiments, the methodology of analysis, and substantive philosophical positions. I hope to achieve this goal *en passant* in the course of arguing for the main thesis: definitions must preserve borderline cases to the same extent as clear cases.

I. DISAGREEMENT AND CONFUSION ABOUT VAGUE DEFINITIONS

Most philosophers regard precision as a desideratum for definition. This value might be defended as a special case of the desiderata operative in theorizing in general. We like out theories to be consistent, complete, general, simple, fruitful — and precise. Observe how astronomers twinkle while telling the story of Kepler's "war on Mars": lesser thinkers would have been content to discount the small variation from his prediction of a circular orbit, but Kepler's high standards of precision led him to scrap the old theory and to discover that Mars' orbit was (slightly) elliptical. He was then poised to generalize to his first law: the orbits of all planets are ellipses with the sun at one focus. Precision pays. Since definition plays an important role in theory construction, one expects it to conform to standards of good science and that means eliminating vagueness.

And indeed, many philosophers discourage vagueness in definition. For instance, Nelson Goodman views "gap filling" as an important function of definition. Some philosophers take precision to be a *basic* desideratum while others trace its value to more fundamental desiderata such as testability. In the case of Keith Lehrer, W. V. Quine, and Peter Unger, the more basic value is consistency because they take vagueness to be a form of inconsistency. Since inconsistency is widely viewed as a severe defect, incoherentists tend to be the biggest pushers of precision. Even those more tolerant of vagueness still impose minimal requirements of precision. For instance, the otherwise easy-going shoulder shrugger William Lycan requires that borderline cases

J.H. Fetzer, D. Shatz and G. Schlesinger (eds.), *Definitions and Definability: Philosophical Perspectives*, 71—109.

be chaperoned by some clear positive cases. Ordinary language philosophers chafed at the excessive vagueness of the animist's definition of 'life', the sceptic's 'doubt', and the materialist's 'physical thing'. They insisted that a predicate is meaningful only if there is a contrast between it applying and not applying. Like Lycan, they would reject the idea of a meaningful but totally vague predicate.

Borderline cases are not well understood and so variations in our subliminal analogies significantly influence rationales for precisification. Many philosophers think borderline propositions are akin to meaningless ones and picture meaningless propositions as semantic holes. So they motivate precisification as a useful precaution — like filling potholes. Others picture borderline regions as undifferentiated semantic matter. The philosopher is free to shape it as he pleases or to just leave it in its formless state. Still others regard borderline propositions as non-functional elements whose weight must be born by the operative remainder. Being vague is like having a dead cylinder or an appendix or "junk genes". Under this analogy, precisification is a sensible reaction to inefficiency. If the thing isn't doing anything, then fix it or get rid of it! But what if its too costly to fix or remove? Well then you might just have to live with the damn thing.

Not all philosophers oppose vagueness. Some even claim that vagueness is a desideratum of definition. Most proponents of truth-conditional semantics take vagueness in a definition to be a virtue — as long as it is coordinated between definiendum and definiens. Others attribute functions to vagueness such as making propositions probable enough to assert. Still others view vagueness as an inevitable side-effect of adaptive devices such as metaphor. In practice, defenses of theoretical definitions seem guilty of special pleading. The same philosopher will at times praise the new precision introduced by a definition and praise the definition for respecting areas of indeterminacy (sometimes quoting Aristotle's principle that one should not demand more precision than one's subject-matter allows).

II. THE NATURE OF BORDERLINE CASES

Although vagueness is in many respects a puzzling and controversial phenomenon, there is considerable common ground amongst commentators. After mapping this zone of agreement, I will try to expand it with some distinctions designed to prevent verbal disputes and expose false paths. I then explore the zone of substantive disagreement to set up an account of the more recalcitrant disputes about definitional policy.

A. *The Consensus about Borderline Cases*

Commentators on the sorites argument agree that a predicate is vague just in case it has borderline cases. All but one group (the incoherentists) agree that a borderline case is one that resists inquiry. If running a car into a body of floodwater is a borderline case of 'collision', then no amount of empirical or conceptual research will demonstrate that it is indeed a collision or indeed not a collision. If pressed for a verdict, we could *stipulate* that it is a collision. Even if we count this as answering the question, note that we must leave the context of discovery and enter that of invention. So one test of borderline status (and thereby vagueness) is the inconceivability of a successful investigation.

A second test of vagueness is sorites embeddability. For instance, the vagueness of 'creeps' is demonstrated by this roll down the slippery slope:

1. A car moving 1 kilometer per hour creeps.
2. If a car moving n kilometers per hour creeps, then car moving $n + 1$ kilometers per hour creeps.

---

3. A car moving 300 kilometers per hour creeps.

Sorites embeddability is a sufficient condition for vagueness and it is virtually a necessary condition. The one exception would be a predicate that had nothing but borderline cases. Although artificial cases can be cooked up, all naturally occurring vague predicates have either clear positive cases or clear negatives. So sorites embeddability is a virtually infallible test for vagueness.

A third test is especially useful for separating vagueness from ambiguity and generality. Here we recast the question in terms of a quantified meaning statement. Utterances of w are ambiguous between x and y if some of them mean x and some of them mean y. So, 'nut' is ambiguous between 'crazy person' and 'shelled seed'. Utterances of w are general if all of them mean (x or y). For instance, 'child' is general between 'boy' and 'girl'. And the utterances are vague between x and y when it is as if they might always mean x or might always mean y but we can never tell. For instance, 'naked' is vague between 'unclothed' and 'uncovered' — thus our indecision about a woman wearing nothing but a net.

B. *Some Worthwhile Distinctions*

Although the following distinctions are not explicitly drawn by commentators on vagueness, they are committed to accepting them. So I take this section to be a further articulation of the consensus.

1. *The Distinction between Hard Cases and Borderline Cases*

The notion of a hard case is more general than that of a borderline case: hard cases are ones that are difficult *or* impossible to settle. The dolphin is a hard case for 'mammal' because it has a striking but ultimately superficial resemblance to big fish such as sharks and cod. However, it is not a borderline case of 'mammal'. Currently, Mars is a hard case for 'sterile planet' but no one regards it as a borderline case of 'sterile planet'. Hardness can be due to *remedial* obscurity.

Overlooking the distinction between borderline and hard cases breeds a bias in favor of precisification. Theories get credit for resolving difficulties, so some thinkers assume that the resolution of any hard case helps a theory. This is too fast. Although a definition earns credit for solving the solvable cases, there is no possibility of discovering the true answer for the classification question posed by a borderline case. So the definition cannot manifest problem solving power by producing answers to the unanswerable questions about these twilight entities.

2. *The Ambiguity of 'Clear Case'*

Although the concept of a hard case is straightforward, 'clear case' is a tricky expression. Note that 'clear' is an excluder word like 'free', 'direct', and 'real'. The excludees for 'clear' in this context are difficulties. Since there are many sorts of difficulties, there are many ways a case can be unclear. However, vagueness theorists mean 'clear case' to merely exclude logically insurmountable difficulties. So they count 541 as a clear case of 'prime number' even thought its primeness is an arcane fact. Most usage of 'clear case' is less extreme. Misleading resemblances, missing stereotypical features, and hidden possession of an essential property suffice to make the case unclear. Thus in ordinary parlance, eels are not clear cases of fish because eels resemble snakes. Penguins are unclear cases of birds because they are among the few exceptions to the stereotype that birds fly. And radio waves are not clear cases of waves because they are invisible. However, the vagueness theorist will not count any of these as borderline cases — they are clear cases in his usage because they are determinable in principle. Also note that irremediability distinguishes borderline cases from near-misses. A woman engaged to a man who dies just before his wedding vow is a near miss of 'widow' because although she fails to satisfy a necessary condition of widowhood, she came close to qualifying.

Once we distinguish between the various ways 'clear case' is used, we can see that the methodological cry for clear cases is not aimed specifically against borderline cases. Often, the request for clear cases arises from a desire for exemplars, paradigms, and trainer cases. This hunger for heuristics underlies D. M. Armstrong's usage of 'clear case' in his frequently cited methodological comment on Colin Radford's "unconfident examinee". Radford's thought experiment is aimed at the belief condition common to nearly all definitions of 'knowledge'. The star of the story is Jean, a French-Canadian who sincerely professes ignorance of English history but agrees to be quizzed on the topic to please a friend. Although Jean takes himself to be guessing, he does enough to exclude the possibility that the accuracy is a matter of luck. As it turns out, Jean eventually recalls taking a class in English history long ago, so the memory must have quietly lingered on. Radford concludes that when Jean was asked when Queen Elizabeth died, Jean knew that she died in 1603 but did not believe that she died in 1603. (Radford also takes this to be a counterexample to 'Knowledge implies confidence' and the KK principle: if you know that *p*, then you know that you know that *p*.) Armstrong's reply is that *if* Jean knows, then he unconciously believes even though he consciously disbelieves. Armstrong's concluding paragraph explains his coyness:

A smaller, but still intensely interesting, question is whether the case that Radford gives is indeed a case of knowledge and belief, or is a case of neither. Unfortunately, however, I do not think that this question can be cleared up until we have a satisfactory account of (a) the nature of knowledge and (b) the nature of belief. For although my intuitions run with Radford's thinking it a case of knowledge (and so, in my view, a case of belief) I do not think that it is one of those clear cases that can be used as a test of philosophical analysis. Rather, we must first develop a theory of the nature of knowledge and of belief, basing it on securer evidence, and then see if our theory will accept

Radford's case. I think that the correct account of knowledge and belief *does* countenance Radford's case, but I can do nothing to prove my point here.[1]

Armstrong does not believe that Radford's case is borderline. For he believes that the question can be answered with the "securer evidence" provided by theory — not a stipulation by the theorist. Armstrong is postponing the question, not rejecting it as indeterminable. His methodological principle of sticking to clear cases does exclude borderline cases as extreme instances of unclarity. But like the lawyer's saying "Hard cases make bad law", the principle is not aimed at vagueness *per se*. The focus in this paper will be on methodological principles that specifically target borderline cases.

3. *Two Senses of 'Vague'*

One verbal dispute can be avoided by acknowledging another sense of 'vague' that means underspecific.[2] A statement is underspecific when it provides insufficient content. For instance, an advertisement for a job paying between \$20 000 and \$200 000 is vague because it leaves an overly wide band of possibilities. Social science terms that lump together disparate common-sense categories are sometimes criticized for this type of vaguification:

> Equal lack of discernment underlies the use of the term 'reinforcement', which in the psychological jargon has replaced the word 'incentive' in the wide sense of the word, which covers positive incentives (that is, rewards) and negative incentives (that is, punishments). As is always the case with jargon, this substitution confuses the issues instead of clarifying them because, whereas 'incentive', 'reward', 'punishment', and 'deterrent' refer only to the manipulation of motivation of some human beings by others, or of higher animals by men, the word 'reinforcement' is much vaguer, as it can be applied to military operations, building techniques or the manner of advancing arguments in a debate.[3]

Notice that borderline cases are not involved here. No sorites can be built from this sort of vagueness. Purely mathematical statements can be underspecific. Also notice that underspecificity has an analytic sort of badness because it is defined as a deficiency of content. Compare 'vague' in this sense with 'undersubscribed', 'undercooked', and 'underfed'.

The ambiguity of 'vague' is relevant because many people extend the vices of underspecificity to borderline cases. They feel that there is something evasive, wishy-washy, or illegitimate about borderline propositions. Some even believe there is legal recognition of the flaw; aren't laws declared "void for vagueness"? However, a borderline proposition can be quite specific: 'Exactly 0% of ambidexterous men are right-handed', 'Chartreuse is yellow rather than green', 'All human beings have tails'. The trouble is that their specificity is useless because borderline propositions cannot be known facts. In summary, vagueness in the underspecificity sense is analytically bad (in the way under-stocked stores are) so an equivocation can lead to a spurious condemnation of the possession of borderline cases.

Conversely, we find commentators confusedly crediting vagueness with the virtues of generality. For instance, writers at least as far back as Pierre Duhem have claimed that vagueness is essential to assertibility because precision and probability are inversely related.[4] This is a spurious justification of vagueness.

'Precision' has a parallel ambiguity. Sometimes it means freedom from borderline cases and sometimes it means high specificity. 'Precisification' inherits this ambiguity. Legislators who precisify 'light beer' so that it means more than 'brand of beer having less calories than another brand produced by the brewery' are increasing the content of the term — they are not eliminating borderline cases.

C. *Rival Theories of Borderline Cases*

The consensus about vagueness breaks down when we try to explain why borderline cases resist inquiry. Epistemic theorists say that the application of a predicate to one of its borderline cases yields a statement with an unknowable truth value.[5] Supervaluationists say that the statement lacks a truth value. Many valued theorists claim that the statement has a degree of truth that is less than the full truth of one but more than the full falsehood of 0. Behavioral theories of vagueness picture predicates as having ranges of indecision where the speech communities' habits are at variance. And there are philosophers, such as Soren Hallden, who call the borderline propositions meaningless. The incoherentists are the one exception to the view that all borderline cases are inquiry resistant. They think that borderline propositions are inconsistent and so can be discovered to be (degenerately) false.

III. HOW DEFINITIONS CAN PRESERVE BORDERLINE CASES

Superficially at least, it seems impossible for a *definition* to preserve vagueness. Indeed, one of H. L. Dreyfus's arguments against linguistic mechanism relies on the tension between the clarity of definition and fuzziness of human concepts.[6] A machine model of human linguistic capacity requires rules expressing the extensions of our predicates. These rules would be biconditionals assigning truth conditions. But, Dreyfus continues, Wittgenstein has demonstrated that the extensions of our predicates cannot be expressed by biconditionals because they are inherently fuzzy-edged. Jerry Fodor objects that this overlooks the possibility of matching the indeterminacies of the definiendum with indeterminacies of the definiens.[7]

Knowing that 'chair' is defined as 'portable seat for one' will not enable me to resolve borderline cases like a "chair" made out of soap bubbles or a packing case that is used as a chair. But that's okay because a good definition need not be a good test. Recognition of this point is manifest amongst contemporary definers of truth. They sharply separate the semantic and epistemological readings of 'What is truth?' An accurate definition can entirely lack verification conditions. Consider Hume's definition of 'miracle' as divine intervention contrary to the laws of nature. It captures the meaning of the term. But, as Hume stressed, the definition also makes the recognition of miracles impossible. Events violating our best established laws are not rationally credible because it is always more likely that the "miracle" is the result of fraud or error. Purveyors of epistemological scepticism about *F*s typically base their position of a definition of '*F*'. Other expressions lacking confirmation conditions include 'telepathy', 'absolutely flat', and 'immortal'. Conversely, a good test need not be a good definition. A good test for whether an egg is hard-boiled is the stability of its spin. For a solid interior causes a smooth, stable spin, while a liquid interior causes wobbles and even continued movement after the egg is suddenly stopped. Nevertheless, a hard-boiled egg cannot be defined as one having a smooth, stable spin.

We are most apt to confuse defining and testing when the term is controversial (art, pornography, religion). For if our motive is to adjudicate, we will crave criteria that deliver a clear verdict. Definitions that coordinate the vagueness of the definiens with that of the definiendum will be poor adjudicators and so stand a substantial chance of being rejected on that score alone. Our goal of conflict resolution will also tempt us to conclude that the vagueness of a term makes if undefinable.

Occasionally we find belief in the incompatibility of vagueness and definition based on the infeasibility of coordination. Ernest Nagel wrote,

> . . . if the word is vague, not only is the range of its application indeterminate; it also is most unlikely that any expression or combination of expressions in the language will convey the customary sense of the word — for the penumbra of indeterminism associated with one set of expressions in the language does not in general coincide with the penumbra associated with another set.[8]

However, there are plenty of examples of successful coordination: 'A puppy is an immature dog', 'Noonish means near noon', 'A middle-aged man is one who is no longer a young man but not yet an old man'. Unless the indeterminacy is ineffable, we should expect that our expressive powers will enable us to tailor the definiens to the definiendum. Any detectible variance in the penumbras of indeterminacy can be corrected with an extra clause in the definition. So Nagel's assessment is too glum.

IV. ARE A TERM'S BORDERLINE CASES PART OF ITS MEANING?

Once it is granted that a definer *can* preserve vagueness, there still remains the question of whether he *ought* to. Your answer is chiefly determined by your stand on the question of whether a term's borderline cases contribute to the term's meaning. Philosophers who think that borderline cases make no semantic contribution see no reason to feed these freeloaders. Those who think borderline cases make up some of the meaning see good reason to preserve them. After all, definers try to preserve as much of the original term's meaning as possible. So if a term's borderline cases are part of its meaning, there is a presumption (albeit overrideable) in favor of preserving vagueness in the definition. The spirit of this conservatism is expressed in Raziel Abelson's criticism of Plato's theory of definition.

> The metaphor of the unseen model also suggests that definitions provide us with precise and rigorous knowledge in the way that blueprints make possible a high degree

of uniformity and precision in productive arts such as architecture. But definitions increase precision only when they change the original meanings of words for technical purposes. Generally speaking, a definition can be no more precise than the concept it defines, at the risk of shifting to a different concept. Our concept of what constitutes an adult is vague; if we try to make it precise by specifying an exact age at which childhood is divided from adulthood, we merely lose sight of what we started out to talk about by replacing the concept of maturity with that of having passed a certain birthday.[9]

The strong version of Abelson's position (which I will eventually defend) is a parity thesis in which the presumption against altering borderline cases is just as strong as the presumption against changing clear cases. A weak version allows a double standard: there is a presumption against altering borderline cases but it is weaker than the presumption against changing clear cases.

Abelson's conservatism presupposes that a term's borderline cases contribute to its meaning. But if the borderline cases are semantic nonentities, there is no reason to keep them around. And if there is some drawback to them (inconsistency, untestability, inefficiency, etc.), definers should eliminate borderline cases. How do we settle the question of whether borderline cases make a semantic contribution? Philosophers agree that the clear cases contribute to the meaning of a term because they constitute the term's extension and counter-extension. But the controversy surrounding the nature of borderline cases ensures that the answer to the question of their semantic status follows partisan lines.

A. *Theories of Vagueness Supporting Enfranchisement*

The epistemic theorists will want to enfranchise borderline cases because they think that borderline propositions have (hidden) truth values and so contribute to the extension of the term just as much as the clear cases. The position has two surprising consequences for synonymy. First, it is possible for two terms to share all of their clear cases without being synonymous. For the breakdown of synonymy could occur beneath the cloud cover of their borderline zones. In addition to implying surprising non-synonymies, the epistemic theory gives us surprising synonymies: every vague predicate is synonymous with some precise predicate in the sense that they have exactly the same truth conditions. Such predicates only differ epistemologically. Consequently, the epistemic theorist's desire to preserve borderline cases is based on the high *risk* that a precisification will alter the meaning of the definiendum. This runs counter to the Abelson intuition that a precisification *inevitably* alters the meaning.

Although the many-valued theory lacks these strange consequences, it also confers semantic status on borderline cases. For it says that clear and borderline propositions only differ in their degree of truth: clear propositions have the extreme truth values of 1 and 0, borderline cases have intermediate truth values. Does a proposition with a 0.8 degree of truth contribute more meaning than one with only 0.7? We are tempted to say yes because we think that one makes a heavier contribution to the *truth* conditions. But falsehood makes an equal contribution. After all, we do not wish to say that a statement can be more meaningful than its negation. Such meditations reveal the many-valued theorist's commitment to the strong parity thesis: borderline statements and clear statements contribute equally to meaning. So he shares the epistemic theorist's egalitarianism.

Incoherence theorists make up the third group of philosophers who think a term's borderline cases contribute to its meaning. However, most of them count inconsistency as a vice that overrides the presumption in favor of preserving meaning. According to this subgroup, vagueness will only be preserved in a reportive definition — where the word is portrayed "warts and all". Explicative definitions will eliminate as much vagueness as possible because inconsistency is a very bad theoretical and administrative vice. A smaller subgroup of incoherence theorists think that inconsistency is not as bad as philosophical tradition assumes and so believe that some definers are justified in leaving the vagueness.

B. *Theories of Vagueness that Ostracize Borderline Cases*

Ostracizers say that borderline cases do not contribute to the meaning of the term. This is obviously the position of theorists who believe that borderline propositions are meaningless. One unwelcome consequence of this harsh position is that borderline propositions never differ in meaning. Yet 'Ghosts are dead' and 'Ghosts are living things' have different meanings. And they certainly differ in meaning from the borderline proposition 'A man standing on his balcony is outside the building'.

One way to cushion the blow is to say that although borderline propositions are alike in that they all lack *actual* meaning, they differ in their *potential* meanings. The idea is that borderline statements provide guidelines for their metamorphosis into fully meaningful statements. Meaningless words such as 'flerp' and 'slarf' do not impose this constraint. 'Ghosts are flerps' and 'Ghosts are slarfs' do not differ in actual or potential meaning because any precisification of 'flerp' is a legitimate precisification of 'slarf'. 'Dead' and 'alive' have different permissible precisifications. Under some of these precisifications 'Ghosts are dead' comes out true and under others, false. This structuring of hypothetical meanings gives borderline statements something close to actual meaning. Borderline statements are akin to people who are disenfranchised for now but how may later acquire the right to vote. They have no direct influence on any election but the possibility of their future enfranchisement indirectly influences current elections. The notion of potential meaning lets us picture the nonsense word as a limiting case in which potential meaning shrinks to zero. Supervaluationists should say that such terms have no meaning rather than mean something empty. This avoids the embarassment of counting all meaningless terms as synonymous with each other. We would like to say that meaningless terms cannot be synonymous with anything because a term can only share meaning if it has a meaning to share.

This scheme lies in the background of the supervaluational approach to vagueness. From its inception, we see Bas van Fraassen motivating the approach with the idea that the logic of our language sets guidelines for its own evolution.[10] Borderline regions are pictured as areas reserved for future conceptual decisions. As we learn more about the costs and benefits of settling borderline cases one way or the other, we can propose progressively more precise definitions. So instead of prematurely settling everything at once, we can proceed gradually, filling in gaps at the propitious moment, making well-informed, piecemeal decisions. A consequence of this attractive view, emphasized by Kit Fine, is that scientific concepts should become progressively more precise.[11] Of course, most predicates will never be completely filled. Classical logic is thus only accurate for the limiting case of a perfectly precise vocabulary. Happily, this inadvertent idealization is a total success at the syntactic level: all the logical tautologies and contradictions are correctly identified. Thus, the supervaluationist can account for the success of classical logic and even advertise itself as a supplement to classical logic rather than a rival.

V. THE OSTRACIZER'S DOCTRINE OF DEFINITION

In any case, supervaluationism has been growing in popularity. Many philosophers explicitly endorse it (Hans Kamp, Robert Stalnaker, Richmond Thomason). The unofficial alignment of others can be seen in the way they craft their informal discussion of vagueness so that it falls within the party line. Many others are fellow travellers. Together they form a school of thought that can be characterized by a platform on the phenomenon of local meaninglessness.

A. *Three Planks of the Platform*

The naturalistic stance toward language provides a warm, moist environment for the notion of local meaninglessness. From this perspective, words are tools fashioned to meet practical needs. Like other tools, words develop only as far as necessary to perform their functions. Since communication requires some uniformity of usage, there will be social pressure to apply the word in certain cases and to withhold it in others. But since social pressure is a scarce resource, the guidance will be incomplete: in addition to the requirements and prohibitions, there will be silence — especially for cases lying in zones of little human traffic. Those interested in the word can describe these ragged patterns of usage and even extrapolate from this data to rules and semantic theories that are intended to explain the linguistic regularities. This extrapolation is a fascinating business. Indeed it sometimes leads us to reverse the priority of data over theory and to adopt a magical view of language in which words have meanings independent of their actual usage. This instance of the tail wagging the dog is the point of one of Branden Matthews's anecedotes. One of his acquaintances was miffed at the confusion over 'people' and 'persons' and complained that "For twenty-five years or more I have kept my eye on the little word *people* and I have yet to find a single American or English author who does not misuse it."[12] The moral is that usage makes the direction of truth-making. When we conjoin the priority of usage over definition with the incompleteness of usage, we come to expect pockets of meaninglessness. As social pressure dissipates, the basis for meaning fades and there is no longer any fact of the matter. For the facts undergirding the meaning of a term are facts about usage. This dribbling away of meaning leads us into borderline cases. Vagueness is pictured as local meaninglessness. A cluster of familiar doctrines about definition emerge from this picture of vagueness.

1. *Semantic Incompleteness*

The first doctrine is *semantic incompleteness*: a vague definiendum will have cases in which there is "no fact of matter" as to whether the term applies. When defining 'food', for instance, we run into the borderline case of coffee. We should not explain our ignorance about whether coffee is food by stressing that it has a *hidden* truth value. For it can only have a truth value if there is corresponding fact about the usage of 'food' that makes the utterance true or false. Our cognitive relation with borderline statements is that of "abnorance", a term designating our relation toward statements that lack truth value.[13] Ignorance holds between us and propositions with truth values unknown to us. Inquirers may reasonably lament their ignorance because knowledge was possible. To lament abnorance is to wish for something that is impossible; if the statement has no truth value, no knowledge of it is possible. The doctrine of semantic incompleteness also tells us *where* inquiry will be limited by abnorance. We can expect the limits to rise where usage trails off. Thus indicators of falling usage mark one important limit to inquiry.

A second lesson is that if borderline cases are pockets of meaninglessness, then definitions must yield clear cases. A definition that made everything a borderline case of the term would fail to assign any meaning and hence not be a definition. Remember that vague terms are supposed to differ from meaningless ones in that they have potential meaning. If all the word's cases are borderline, it lacks the contours constitutive of potential meaning and so is entirely meaningless, not vague. This reasoning may lie behind William Lycan's requirement that all vague predicates have some clear positive case.[14]

In addition to these methodological points, we find semantic incompleteness used to support substantive claims. For example, most of us believe that there is always a yes or no answer to the question of 'Am I about to die?' This conviction supports the idea of a Cartesian ego (when supplemented with the results of certain thought experiments). Parfit challenges the determinancy of your personal survival with an analogy.

Suppose that a certain club exists for several years, holding regular meetings. The meetings then cease. Some years later, some of the members of this club form a club with the same name, and the same rules. We ask: "Have these people reconvened the *very same* club? Or have they merely stated up *another* club, which is exactly similar?" There might be an answer to this question. The original club might have had a rule explaining how, after such a period of non-existence, it could be reconvened. Or it might have had a rule preventing this. But suppose that there is no such rule, and no legal facts, supporting either answer to our question. And suppose that the people involved, if they asked our question, would not give it an answer. There would then be no answer to our question. The claim "This is the same club" would be *neither true nor false*.

Though there is no answer to our question, there may be nothing that we do not know. . . . When this is true of some question, I call this question *empty*.[15]

Parfit's analogy is intended to make a metaphysical point. Just as there is no hidden fact that makes the club continue to exist or not exist, there need be no hidden fact determining whether you continue to exist.

2. *Semantic Pragmatism*

The search for indicators of falling usage leads to the second doctrine of *semantic pragmatism*: usage (and therefore meaning) is driven by the needs of the linguistic community. So our vocabulary only has as much meaning as is necessary to deal with communication problems. Words will fail us when we confront abnormal situations. Suppose you meet a two headed man and cut one of his heads off. Would you have decapitated him? We feel torn between rival rules. 'Decapitate' might mean cut a head off or it might mean make headless or cut *the* head off or even mean cutting the most important head off. Our usage does not decide the issue because it never had to. If two-headedness became common, our language might well fill the gap. But as the language actually stands, there is no fact of the matter. Since any appeal to "what we would say" must be relativized to our actual language, semantic pragmatists are sceptical of thought experiments involving bizarre suppositions. W. V. Quine expresses this caution in his criticism of the debate between Sydney Shoemaker and David Wiggins over personal identity:

Later [Shoemaker] examines Wiggins on personal identity, where the reasoning veers off in familiar fashion into speculation on what we might say in absurd situations of cloning and transplanting. The method of science fiction has its uses in philosophy, but at points in the Shoemaker-Wiggins exchange and elsewhere I wonder whether the limits of the method are properly heeded. To seek what is "logically required" for sameness of person under unprecedented circumstances is to suggest that words have some logical force beyond what our past needs have invested them with.[16]

Many philosophers take Quine's scepticism to heart and become leery of outlandish thought experiments even when these stories spawn firm beliefs. For they take Quine's key point to be that we lack a *justified* belief about what to say in these alien conditions, not that we have no beliefs. Note that the scepticism would also extend to some non-hypothetical cases. Since dinosaurs predate humans, we never had to settle whether the *Archaeopterx* was a bird rather than a reptile. In addition to ignoring temporally remote cases, we also draw a blank for culturally remote ones. Speakers of English live in monagamous cultures and so are at a loss when asked 'If a woman has two husbands and one of them dies, is she a widow?' They do not know whether to define a widow as a woman with a dead husband who has not remarried (in which case the polyandrous woman is a widow) or as a woman rendered unmarried by the death of her husband (in which case the polyandrous woman does not qualify as a widow). The fact that this situation does arise in a few contemporary societies does not alter the indeterminacy. Rarity and novelty will also produce real-life borderline cases. Do Siamese twins remain Siamese twins after surgical separation?

3. *Semantic Discretionism*

A related lesson about thought experiments flows from the third doctrine of *semantic discretionism*. If I define a term in a way that runs contrary to ordinary usage, then either my definition is inadequate or I assume the burden of justifying this stipulative aspect of my definition. So there is a presumption against definitions that conflict with clear cases. (A conflict here is changing a clear positive to a clear negative, a clear negative to a clear positive, or a clear case to a borderline case.) Discretionism says that there is no corresponding presumption against conflict with borderline cases. The definer is free to fill in the gaps as he pleases. Nelson Goodman portrays this gap-filling as a welcome byproduct of definition:

. . . common usage is indeterminate with respect to certain entities. The term "fern", for example, we unhesitatingly apply to certain plants and refuse to apply to others; but there are some plants about which we are undecided, not because we lack knowledge of these plants but because habits of using the term "fern" are not fully determinate. When any definition meeting this requirement is established, it will then serve as a means for classifying the hitherto borderline cases (e.g., as ferns or non-ferns). Thus a definition accepted on the ground that it violates no manifest decision of ordinary usage becomes legislative for instances where usage does not decide.[17]

One sign of semantic discretionism's influence is that definers who leave indeterminacies are apologetic — reminiscent of hard-pressed custodians who cannot finish sweeping. Lawrence Becker, for example, concedes that his definition of 'human being' does not provide answers to bizarre questions such as whether a disembodied but functioning brain is a human being. But he defends the omission by saying that one could only get to those hard to reach spots by damaging the definition in other respects:

> . . . the inability of a definition to settle bizarre cases need not be considered an overwhelming defect. Bizarre cases can often be settled only by equally bizarre definitions. The definitions proposed here — for both the becoming/being and the being/has-been boundaries — make good sense conceptually, are sufficiently clear for moral purposes, and direct our attention to the moral issues surrounding homicide in a productively direct way. That much, it seems to me, is enough to expect from definitions.[18]

Becker agrees that it would be *nice* to eliminate the borderline cases. He just denies that this kind of completeness is obligatory. Just as an almost completely clean room can be clean enough, an almost precise definition can be precise enough.

One interesting consequence of semantic discretionism is that no definition is even slightly disconfirmed by a borderline case. Indeed, gap-filling is praised. (Correspondingly, gap-making is blameworthy, a more severe evil than the sin of omission: gap-leaving). Hence a thought experiment or apparent counterexample can be disarmed by showing that the scenario is only a borderline case of the predicate under analysis. Also note that a definer can secure greater and greater liberty by showing that apparently clear cases are really borderline. Thus ostracizers treasure arguments that have this liberating effect — ones that unmask "false determinacies". Definers become fond of emphasizing the difference between clear cases and ones that are merely familiar, desirable, or widely accepted. They are also on the prowl for doctrinal biases — and not just those of intellectuals. Consider the widespread conviction that a person ceases to exist upon his death. Is a corpse really a *clear* negative case of a person? Why isn't a corpse just an irreversibly insentient person? Perhaps the judgement of even common folk has been biased in favor of sentience by religious doctrine.

Some philosophers try to extract substantial philosophical consequences from the liberty bestowed by unsuspected vagueness. Proponents of the "brain death" criterion agree that a person in an irreversible coma is a borderline case of a dead person. Traditional bodily criteria, codified in law, count the patient as alive because his blood is circulating. But the patient does not satisfy another intuitive requirement: having mental capacities. So we are faced with competing definitions of 'life'. Some people suggest that we are free to break the deadlock by appealing to the legal and medical benefits of counting the coma victim as dead. Jonathan Glover puts the matter more broadly:

> The only way of choosing (between competing definitions of death) is to decide whether or not we attach any value to the preservation of someone irreversibly comatose. Do we value 'life' even if unconscious, or do we value life only as a vehicle for consciousness?[19]

Michael Green and Daniel Wikler criticize this position by drawing attention to the absurdity of reclassifying horribly handicapped people as dead merely on the grounds that they are better off dead.[20] However, Glover's position can be resuscitated with the observation that the horrible handicapped are not *borderline* cases of 'dead' and so lie outside the scope of semantic discretionism.

A more aggressive application of semantic discretionism can be seen in an idea floated by Mark Johnston in a series of lectures at New York University. Johnston speculates that the surprising indeterminacies discovered by personal identity theorists support a theme running from the Renaissance to the existentialists: man is a self-defined being. Nietzsche captures the spirit of self-made man when he writes "We, however, *want to become those we are* — human beings who are new, unique, incomparable, who give themselves laws, who create themselves."[21] The borderline cases for 'person' constitute semantic prime matter that can be molded as we see fit. Unlike other theorists who would use only linguistic considerations to fill in the gaps, Johnston thinks we are entitled to use criteria which are more frankly prudential. A rationale for this might be that words are ultimately there to serve our interests, so we can skip the intermediary steps and appeal directly to what is in our interests. So given a choice between rival precisifications of 'person', we are free to adopt the one that maximizes the value of our lives. As an illustration, Johnston has us suppose that we develop Star Trek-style teletransportation but then wonder whether we really survive the "trip". Maybe the departees are only destroyed in one place and duplicated at the other. Johnston denies that there is a fact of the matter because the scenario features a borderline case of 'personal survival'. However, we can precisify 'personal survival' in a way that ensures that we do survive. By doing so we would secure the advantages of rapid transportation (or if that isn't enough, further suppose that the process also eliminates the diseases of the departee). Given these advantages, Johnston thinks we would be wise to hotwire 'person' to allow survival through spatial discontinuity and loss of one's original matter.[22]

Semantic discretionism is not confined to logicans or philosophers. It is also popular amongst linguists partly because it was endorsed by Noam Chomsky in the opening chapter of his seminal *Syntactic Structures*. However, linguists have complained about discretionism's limited feasibility. One common objection is that cases over which linguists have clashing intuitions tend to be just the ones their rival theories disagree about. When two theories align with the diverging intuitions, application of discretionism by either side is question-begging. A further impracticality looms when we consider the linguist's two favorite explanations of speaker divergence: neither of the standard moves explains the divergence in terms of *vagueness*. The first explanation appeals to the competence/performance distinction: at least one of the intuitions is caused by a factor interfering with performance, such as inattention or bias. For instance, if most women affirmed 'Geldings are stallions' while most men denied it, we would suspect distorted judgement. In principle, one could correct these failures, so this kind of unclear case is not a *borderline* case. Therefore, semantic discretionism fails to apply. The second explanation is that the speakers vary in their internalized grammars. But this explanation makes vagueness equally irrelevant. For linguists are supposed to *describe* usage rather than *prescribe* it. Since description of usage amounts to codification of internalized grammars, the linguist must conclude that relative to one grammar the disputed case is to be settled one way, and relative to the other grammar, another way — which is to say there really is no disagreement between the "conflicting speakers". The speakers' dispute was merely verbal because each was unwittingly relativizing their utterances to different internal dialects. Since relativity differs from vagueness, the doctrine of semantic discretionism does not apply.

The linguist's qualms about application are sound but their impact is easily overestimated. To see why, we must distinguish between two conceptions of a borderline case. The neutral conception pictures well-informed speakers as *lacking* an opinion when asked about a borderline case. They just shrug their shoulders when asked whether sponge is a substance rather than an artifact. The partisan conception pictures well-informed speakers as having *conflicting* opinions when asked about a borderline case: some speakers insist that sponge is a substance while others are sure it is an artifact. Yet other speakers may vacillate between the substance and artifact answers. And perhaps there are some who hold both opinions simultaneuosly and so are inconsistent. Some semantic discretionists have this partisan conception in mind. For example, Katz and Fodor have said that anything over which speakers disagree is *ipso facto* not part of linguistic data.[23] Thus a partisan borderline case is never evidence for or against a linguistic hypothesis. However, discretionists generally have the neutral conception in mind — thus the language of gaps, fading usage, and "don't cares". Now notice that the feasibility criticism of the previous paragraph is targetted against *partisan* borderline cases — notice the language of conflict: clashing intuitions, disagreement, rival theories. A semantic discretionist could concede that there are feasibility problems with the partisan cases while insisting that the discretionism is practical for the kind of borderline cases he was studying. Alternatively, he could try to undermine the partisan conception by denying that well-informed and competent speakers can have that sort of disagreement.

VI. IS THE PARTISAN CONCEPTION COHERENT?

Although the neutral conception of borderline cases has been the dominant one, the partisan conception has played an important role in thought about vagueness and definition. We have just seen how it is relevant to semantic discretionism and we will see it is of more central significance to the behavioral and incoherentist theories of vagueness.

A. *Proponents of the Partisan Conception*

The partisan conception of a borderline case lies behind one of the earliest attempts to quantify vagueness. Hempel defined vagueness in terms of speaker variation.[24] Given a symbol *T* and an object *x*, we ask members of a group 'Does *T* apply to *x*?'. This provides us with a ratio.

$$\frac{\text{The number of yes-answers}}{\text{The number of questions}}$$

Hempel then defined the consistency of application, $C(T, x)$, to be the limit of the ratio as the group of observers and questions is indefinitely extended. Clear cases will be those in which $C(T, x)$ is close to one or zero. Borderline cases will be those close to 1/2.

The partisan conception of vagueness was not confined to logical positivists or behaviorists. Many passages from Wittgenstein's writings suggest that linguistic communities are divided in ways that make some statements indeterminate.

> There is such a thing as colour-blindness and there are ways of establishing it. There is in general complete agreement in the judgements of colours made by those who have been found normal. This characterizes the concept of a judgement of colour.
>
> There is in general no such agreement over the question of whether an expression of feeling is genuine or not. I am sure, *sure*, that he is not pretending; but a third person is not. Can I always convince him? And if not is there some mistake in his reasoning or observations? 'You don't understand anything!' — we say this when someone doubts what we recognize as clearly genuine — but we cannot prove anything.[25]

Norman Malcolm cites this passage when constructing a rival explanation of the evidence Daniel Dennett cites for the thesis that 'pain' is inconsistent. Dennett points out the same textbook will say that nitrous oxide renders one insensible to pain and then later classify nitrous oxide as an analgesic — a preventer of pain.[26] A second example centers on talk of pain thresholds. How can we distinguish between a person who has great toleration of pain and one who simply feels less pain? The same question arises for reports of prefrontal lobotomy patients who say that although they still feel the pain it bothers them less. Dennett writes,

> The grammatical grounds for the contradiction have already been noted: it is equally ordinary to speak of drugs that prevent pains or cause them to cease, and to speak of drugs that render one insensitive to the pains that may persist. . . . So ordinary usage provides support for the view that for pains, *esse est percipi*, and for the view that pains can occur unperceived. . . . What must be impeached is our ordinary concept of pain. A better concept is called for . . .[27]

Dennett applies this thesis to obtain a nihilist answer to the question of whether a computer could feel pain.

If, as I have claimed, the intuitions we would have to honor were we to honor them all do not form a consistent set, there can be no true theory of pain, and so no computer or robot could instantiate the true theory of pain, which it would have to do to feel real pain.[28]

Malcolm's rival explanation is that 'pain' has indeterminate cases. He begins his argument by adding some further troublesome situations. Suppose that in a walk with a friend, your legs begin to ache. But you and he begin a lively conversation and you no longer notice the ache. When the conversation ends, you again become aware of your ache. Did the ache continue? In *Captain Horatio Hornblower*, C. S. Forester writes, "Hornblower found the keen wind so delicious that he was unconscious of the pain the hailstones caused him." Could this attribution of unfelt pain be true? Malcolm also has us consider pain-threshold cases where severely wounded people only complain slightly and where slightly wounded people complain severely. He notes that there is a parallel doubt concerning the expression of emotions and attitudes. Wittgenstein commented on further puzzle cases in *Zettel.* Wittgenstein reports that some people confidently say that general anaesthesia makes you feel nothing while others say "It *could* be that one feels something but forgets it completely."[29] Wittgenstein further notes,

We often use the phrase 'I don't know' in a queer way; when for example we say that we don't know whether this man really feels more than that other, or merely gives stronger expression to his feeling. It is in that case not clear what sort of investigation could settle the question.[30]

Malcolm takes himself to be agreeing with Wittgenstein when he says that these indeterminate situations foment irresolvable disagreement between competent speakers of English. Part of the disagreement could be due to temperamental differences. Trusting folk may be willing to take people's expressions of feeling at face value. Suspicious people will be more inclined to wonder whether people malinger and whether others put on a brave front.

Malcolm's allusion to temperament reminds us of William James' distinction between tough-minded and tender-minded philosophers who are attracted to different positions because of their personalities. Malcolm's talk of irresolvable disagreement also invites comparison with Gallie's notion of "essentially contested concepts" which are "concepts the proper use of which inevitably involves endless disputes about their proper uses on the part of their users."[31] Gallie's examples include 'religion', 'culture', 'artwork', and 'democracy'. Perhaps some philosophical disputes boil down to diverging reactions to borderline cases. The dispute over the possibility of masochism is a promising candidate. The report 'I enjoy some of my pains for their own sake' appears contradictory because 'pain' seems to mean a certain kind of disliked sensation. Perhaps the "masochist" *extrinsically* likes the pain because it brings attention or relief from guilt or whatnot. Trusting types are ready to take the report at face value and so are willing to revise the simple theory of pain in order to accommodate the weight they attach to testimony about one's preferences. Psychological egoism and scepticism about weakness of will might be analyzed the same way.

Partisan borderline cases are naturally pictured as the result of clashing rules. Like some ill-coordinated bureaucracy with overlapping legislative bodies, language may get its lines of authority tangled. However, Malcolm spurns this picture in favor of one which gives priority to agreement rather than rules:

> It might be that the disagreement that exists is to be explained by the absence of clear-cut *rules* for making attributions of feelings to others. It is true that there are no clear rules. But this is not an explanation of the disagreement. Instead it is the other way round. There can be clear, precise rules only when there is virtually universal agreement. The certainty of arithmetic, of judgements of measurement, of the relationships between colours, depend on this. The firmness and clarity of rules is a reflection of overwhelming agreement. That kind of agreement is not present in regard to people's sensations and feelings.[32]

Malcolm goes on to say that we can imagine a society in which there was this sort of agreement about sensations. In this imaginary society, judgements about expressions of feelings would be determinate. Of course, Malcolm is not saying that all uncertainties about feigning are irresolvable. We can resolve some of the disagreements with appeals to past or subsequent actions.

> What I mean, however, by 'indeterminacy' is that there are frequent examples of expressions of feeling, where equally sensitive and informed observers differ in their assessment of the sincerity of the expressions, and where no one can see any way of resolving the disagreement. This means that here there is no definite right or wrong in the matter. One person in such a dispute may say 'You are too gullible', and the other may say 'You are too insensitive', but this is only scolding. Each of the disputing persons may continue to say 'I'm right and you are wrong', but the sense of such remarks is different here from what it is in contexts of arithmetical calculation or measurements of length.[33]

The partisan conception of borderline cases continues to influence most commentators on vagueness.[34] But though alive, it is not well.

B. *The Instability of the Partisan Conception of Borderline Cases*

Malcolm does not specify the difference in sense between a disagreement over a borderline case and a disagreement of a calculation. But some sort of qualification is needed to account for the peculiar nature of the disagreement over a borderline case. When I believe something, I believe it is true, hence the oddity of G. E. Moore's 'It is raining but I do not believe it'. Now note that the same oddity is encountered with 'John is a borderline case of a tall man but I believe he is a tall man' or closer to home, 'it is indeterminate whether Mary is in pain but I believe she is in pain'. There is nothing odd in believing a borderline case of an *F* to be an *F* if you mistakenly think it is a clear *F*. But once you recognize that *x* is a borderline *F*, you no longer believe *x* is an *F*. Asking a person for an example of a borderline *F* that he believes to be an *F* is like asking him for an example of a non-*F* that he believes to be an *F*. Also note the Moorean quality of 'Mary is a clear case of thin woman but I believe she is a borderline case of a thin woman'.

These considerations suggest that partisan borderline cases tend to collapse into neutral borderline cases. For once the disagreement is recognized, we will also recognize that neither side's belief is true (or can be regarded as true). So the effect of the revelation should be dissolution of the disagreement. Well-informed speakers withdraw into agnosticism about the case.

The same instability is displayed by intra-personal versions of partisan borderline cases. A speaker can have a diachronic disagreement with himself in that his linguistic intuitions may vary over time. A boy who counts the family dog as a family member may exclude it when he matures into a grouch. The synchronic case is a special instance of inconsistent belief. For here, one simultaneously believes of a borderline *F*, that it is an *F* and believes that it is not an *F*. Inconsistent beliefs are usually thought to be bad because they guarantee error. But when you are inconsistent over a statement that you consider to be borderline, you are only doomed to error if you think it has hidden truth value. Since nearly all vagueness theorists reject this, they need to explain why it is irrational to both believe and disbelieve a borderline proposition. In any case, the collapse of partisan borderline cases into neutral ones holds for the special case of one person "disagreeing" with himself. As soon as he detects the nature of the disagreement, he suspends judgement.

VII. THE INCOHERENCE OF THE INCOHERENTIST'S DOCTRINE OF DEFINITION

The instability of partisan borderline cases is bad news for the incoherentist's theory of vagueness. Recall that they take borderline propositions to be inconsistent. The partisan conception lends support to this view because clashing speakers can be pictured as elaborating different strands of the inconsistency. According to the incoherentists, the group that believes vampires are dead and the group that believes vampires are alive are both right but incomplete. The only ones who completely capture the meaning of 'vampire' are those who forthrightly affirm both parts of the inconsistency because the term means a kind of blood-sucking creature of the living dead. Thus the price of complete meaning preservation is inconsistency. Since incoherentists agree with other vagueness theorists that most of our vocabulary is vague, they believe that most our vocabulary is inconsistent. Keith Lehrer writes,

> To say that a case is borderline means there are quite conclusive considerations in favour of applying the term, and equally conclusive considerations in favour of not applying it. For example, if we see something that is very similar in colour to many red things, so much so that this is a quite conclusive reason for saying that the object is red; but at the same time it is very similar to things that are orange and not red, so much so much so that this is a quite conclusive reason for saying that the thing is not red, then we have borderline case of something red. Such cases abound. For most terms of everyday speech, we can expect to find that the term applies without doubt or controversy in a large number of cases, and that it also clearly fails to apply in many cases, but in between these cases there are examples of things where it is not evident whether or not the term applies no matter how much one knows about the example.[35]

The widespread inconsistency of ordinary language is a popular theme of recent critics of folk psychology. According to them, our ordinary concepts of personhood and belief "fracture" in the face of phenomena reported in split-brain research and studies of abnormal psychology. Even physical concepts get stressed to the breaking point. Recall how turn of the century problems with 'solid' led Eddington to contrast ordinary tables with scientific tables. One necessary condition of being solid is being entirely filled in, without gaps. One sufficient condition is being constituted like this table and other paradigm cases which 'solid' was taught. Atomic physics disclosed a conflict because things like this table are mostly empty space. So tables count as nonsolids by the necessary condition and solid by the sufficient condition.

A. *Pessimistic Incoherentists*

Since inconsistency is widely regarded as a serious vice, most incoherentists think they have an alarming message with important methodological consequences. For if the price of preserving ordinary meaning is inconsistency, we should conclude that the price is too high and approach folk predicates with scalpel drawn. The only exception to 'When in doubt, cut it out' will be purely reportive definitions where meaning preservation is the only thing that matters. Otherwise, we should expect an adequate definition of a vague term to cut deeply into the original meaning. In the eyes of the incoherentist, the problem is a meaning glut, not a meaning gap. The incoherentist does think that there is a presumption in favor of preserving meaning and does believe borderline cases contribute meaning. Nevertheless, his resulting definitional behavior is much like the semantic discretionist's because the vice of inconsistency overrides the presumption in favor of preserving meaning.

Pessimistic incoherentists are aggressive explicators and so fall into disputes with ordinary language philosophers. The strategy of resolving philosophical problems by careful study of ordinary usage presupposes linguistic orderliness and coherency. But if the rules of ordinary language are confused and ridden with cancerous inconsistency, then the rules must be reformed rather than followed more scrupulously. So the revisionism of incoherentists sharply contrasts with the conservatism of the ordinary language philosopher. For instance, Colin Radford frequently complains that attempts to defend 'Knowledge implies belief' have been more heavily influenced by theory than by careful consideration of how 'know' is actually used. Radford insists that distortions of ordinary meaning can only be justified by substantial benefits.[36] Lehrer deems this squeamish:

> . . . any coherent theory that does not lead to contradictory conclusions concerning the borderline cases we have been considering *must* 'distort' our inconsistent ordinary usage of the terms 'know' and 'believe.' For most purposes of everyday speech we can afford the semantic imperfection that yields the sort of contradictory conclusions we

have derived from the study of Jean and company, but an acceptable *theory* must eliminate such imperfection for the sake of consistency.[37]

Happily, the process of explication is not restricted to the grim business of excising inconsistency. While you are in there, you can make positive improvements because you can precisify in the direction of desiderata such as testability, simplicity, completeness, and fruitfulness. For instance, Lehrer analyzes 'know' in a way that maximizes its implications so that he emerges from his analysis with many pretty principles such as KK and 'Knowledge implies belief'.

The pessimistic incoherentist is no knee-jerk explicator. Often, he just abandons the term because it lacks salvage value. For instance, 'vampire' is a rather peripheral member of our vocabulary, so we might just drop the term — at least for official purposes. Other words, such as 'solid', play a more central role and so are worth tidying up. However, some incoherentists also think that several central concepts (cause, belief, meaning) are "totaled". If pessimistic incoherentism is combined with pessimism about the power of explication to save our vocabulary, we get the "Burn, baby, burn!" nihilism of Peter Unger and Samuel Wheeler. They end up denying the existence of every ordinary thing — in the case of Unger, even himself.[38]

B. *Optimistic Incoherentism*

A minority of philosophers think that the evil of inconsistency has been greatly exaggerated. Although this groups extends back to Nietzsche and Hans Vaihinger, it is enjoying a revival amongst logicians developing paraconsistent or relevance logics — as prophesized by Wittgenstein:

Indeed, even at this stage, I predict a time when there will be mathematical investigations of calculi containing contradictions, and people will actually be proud of having emancipated themselves from consistency.[39]

These deviant logicians attempt to block the familiar proof that a contradiction implies everything by rejecting classical rules of inference such as disjunctive syllogism. Whereas the classicist thinks that a contradiction causes all hell to break loose, the relevance logician thinks that it causes only a little hell to break loose.

Some incoherentists (such as Bertil Rolf) overlap with this minority of philosophers who find inconsistency tolerable. So unlike their pessimistic cousins, these optimists deny that the inconsistency of borderline cases generally overrides the presumption in favor of preserving meaning. That is to say, definers should preserve borderline cases even though this loyalty to meaning engenders inconsistency.

C. *Objections to Incoherentism*

Incoherentism is preposterous. Since only consistent predicates can have instances, the thesis that borderline propositions are inconsistent implies that all vague predicates are empty. But since nearly all of our predicates are vague, incoherentism is committed to repudiating both common sense and science. Nevertheless, it is seriously defended by a large group of eminent philosophers who will not be stared off the podium. So my only follow-up is to indicate how the incoherentism fails in a myriad of specific, technical ways. So let's put away the big picture and turn to some embarrassing snapshots.

First, take a hard look at how the collapse of partisan borderline cases into neutral ones dries up the rationale for picturing borderline cases as inconsistent. The principle of total evidence ensures that conflicting trends cancel out. So loyalty to actual language does not force us to break our teeth on the "hard truth" of illogicality. Second, any attempt to take the characterization of inconsistency seriously will require repudiation of whole blocks of logical intuitions. If borderline statements were inconsistent, they would imply everything because anything follows from a contradiction. But the following argument is invalid rather than trivially valid:

1. Wall-to-wall carpeting is furniture.

---

2. Wall-to-wall carpeting slows down cockroaches.

Likewise, 'If coffee is food, then coffee is a deadly poison' is not made trivially true by the inconsistency of its antecedent. Any attempt to dodge these consequences by rescripting logic will only play to a tiny audience. Most philosophers will either accuse the deviant of throwing away good theory to save bad theory or of just changing the topic (what does 'inconsistent' mean after the rules of inference are refigured?).

Turn next to the incoherentists' commitment to the compositionality of vagueness. (They are committed because conjoining an inconsistent predicate with any other predicate yields an inconsistent predicate.)

There are flat counterexamples to compositionality: the conjunction of the vague 'integer somewhat greater than 104' and the vague 'integer somewhat less than 106' is not itself vague. Fourth, incoherentism violates the principle that the negation of a borderline proposition is itself a borderline proposition. For if the borderline 'My ice cream van is a place of sale' is inconsistent, then its negation is a tautology and so is not a borderline proposition.

Lastly, attributing inconsistency to language re-opens an old can of worms. Recall the objections prompted by Tarski's claim that all natural languages are inconsistent. These criticisms will be revived and amplified against the incoherentist because he attributes inconsistency to nearly all predicates rather than to the handful associated with 'true'. Critics will object that language is the *means* by which we assert, it does not itself assert anything, so language cannot have a thesis, let alone a contradictory one. It is one thing to have predicates that are impossible to instantiate, quite another for them to render language itself inconsistent. The *definition* of 'round square' is perfectly consistent; adding it to the axioms of geometry only leads to the theorem that there can be no round squares. Attributing inconsistency to language also runs foul of the principle of charity. This principle tells us to minimize the irrationalities of interpretees but if their language is suffused with inconsistency, its speakers are stuffed to the gills with irrationalities.

VIII. THE CASE FOR PRESERVING BORDERLINE CASES

I have already discussed how the epistemic and many-valued theories of vagueness imply that a term's borderline cases contribute to its meaning. The result is a theory of definition that is conservative with respect to vagueness. All other things equal, definitions should preserve intuitive verdicts of both clear and borderline cases. Thus the epistemic and many-valued theorists are committed to rejecting semantic discretionism. They deny that there is a gap to be filled — something is already in there. Of course, this presumption against precisification can be overridden but only on grounds similar to ones we have for revising clear cases.

A. *An Appeal to Actual Practice*

Conservatism best fits actual definitional practice. Theoreticians discussing definition tend to be bold in planning but timid in execution. The theorist is determined to be the master of words, not their slave, and so resolves to bend the words to his will, heedless of their howls. But when he gets down to cases, the definer is much more cautious. Part of the reason is terminological. If the definer displays complete disregard for the usual meaning of the term, he will be (correctly) perceived as merely stipulating a new sense for an old word. Defining 'child' as 'a clothed poodle' will license some initially surprising but ultimately irrelevant claims about children. To avoid this sort of irrelevance, the definer must claim that there are *important* connections with the old meaning. Mere overlap of truths does not suffice — 'All children are mammals' comes out true under my silly definition as well as any adequate one. Nor is a shared topic sufficient. People regularly distinguish between the ordinary and technical senses of words such as 'resident', 'ball', and 'work'. A failure to capture an important connection will merely lead speakers to treat the definition as a related sense.

Definitions that deliberately eliminate borderline cases are called precisifying definitions. If I offer a million dollars for the fattest kitten in the world, then there is a strong administrative justification for preferring the artificial 'A kitten is a cat younger than six months' over 'A kitten is an immature cat'. Even so, people will only treat the precise version as a temporary technical definition that sacrifices some ordinary meaning for sake of social tranquility.

B. *Other Meaning-Preserving Practices Preserve Borderline Cases*

Indirect quotation, translation, and exegesis also take meaning preservation as a desideratum. But precisification has little acceptability in these quarters. When asked to paraphrase a speaker, you are expected to leave the vagueness. Failure to do so constitutes a distortion of meaning. This requirement of vagueness preservation suggests that a term's borderline cases do contribute to its meaning.

Latitude is given to those who "rationally reconstruct" a position or those who offer an interpretation of a poem or speech. But once we leave these liberalized variants, we find that most meaning-preserving practices are conservative. Definition has more room for precisification because most definitions have meaning preservation as only one amongst many desiderata (such as brevity, testability, and terminological integration). Only purely reportive definitions avoid compromises.

Purely stipulative definitions offer even greater serenity because they abandon meaning preservation altogether. Why, then, are they so rare?

C. *The Inertia of Convention Forces Conservatism*

The reason is that conventions are difficult to implement and grow in difficulty as the shift in convention grows larger. The history of measurement units, spelling reform, and currency contains many surprising illustrations of human recalcitrance. Those who wish anecedotal evidence to be their guide should try introducing a neologism into English.

The problem isn't entirely a matter of brute intransigence. Several advantages accrue from preserving meaning. The less we vary from ordinary usage, the lower the adjustment costs. The transition to the new usage is cheaper if the distance from old to new is short. Since the past cannot be changed, we face some inevitable meaning variance. This is a loss because the variance hampers our use of the past discourse — an especially important consideration since the growth of writing. Thus minimizing the variance reduces the expected amount of misunderstanding.

All of this forces revisionary definers to think small. They can hope to effect a slight change in usage amongst a subgroup of the linguistic community but have little prospect of altering speech in a big way.

D. *Borderline Regions Are Structured*

There are several reasons for thinking that borderline regions are structured, that they are not semantic prime matter that can be fashioned any way we please.

1. *The Lean of a Borderline Case*

First, many borderline cases have a direction and a degree. We lean toward counting some borderline *F*s as *F*s rather than non-*F*s without actually believing they are *F*s. A man who has recently become a borderline case of a fat man leans toward non-fatness but eventually leans toward fatness after further weight gains put him near clear fat men. We also lean in degrees — there's being close to fat and being very close to fat. Once we distinguish between the direction and degree of a borderliner's lean, we develop preferences about precisifications.

We discourage precisifications that "go against the grain" by changing a positively leaning case into a clear negative case. The intensity of our preferences follows the degree of lean.

The lean of a borderline case affects conditionals and arguments. We assent to 'If 12:28 is noonish, then 12:18 is noonish' because it strikes us as tautologous. But its converse does not. Next consider our reactions to the following moral safety argument against abortion.

1. The fetus is a borderline person.
2. One should not risk killing a person.

---

3. One should refrain from abortion.

Most pro-abortionists think that a fertilized egg is clearly not a person but a fetus in its ninth month clearly is a person. Between these clear extremes is a borderline area which is the target of the moral safety argument. Most people do not find the argument persuasive when it is applied to the early end of this borderline zone but do think we should give late late borderline cases the benefit of the doubt. It is as if they assigned the late fetuses a substantial chance of being people and the earlier cases much lower chances. This would subsume the liberal position on abortion under our general policy toward putting others at risk: we are willing to run a tiny risk of killing innocent strangers (while driving, flying, or building skyscrapers) but regard the imposition of a *high* risk of death as wicked.

Many-valued logic readily reflects this differentiation of borderline cases. Whereas the supervaluationist only has the on/off device of the truth-value gap, the many-valued logician can help himself to infinitely many degrees of truth between 1 (for full truth) and 0 (for full falsehood) neatly accounting for direction and lean. Borderline propositions with an already high degree of truth only require a little nudge to make them clear positive cases. But making them clear negatives requires almost as much reform as changing a clear positive to a clear negative. One noteworthy consequence is that gap-making is better than clear case reversal; there is less change from a clear case to a borderline case than from a clear positive to a clear negative (or vice versa).

The epistemic theorist reaches a position similar to the many-valued theorist but by a different route. His conservatism flows from a desire to minimize expected meaning change. A borderline *F* leans toward *F*-ness in proportion to the probability that it really is an *F*. (So a utilitarian epistemic theorist would prefer the death of one clear case of a person over the deaths of two borderline people who leaned toward personhood.) If forced to precisify, we should go in the direction of the lean because this gives us the better chance of preserving meaning.

Those who stress that borderline cases are a matter for a decision may try to explain the lean and degree of borderline cases as an instance of projection: speakers lean, not their predicates. Consider how deliberators are described as leaning toward one option rather than another. The agent has yet to make a choice. So the force of describing him as leaning in the direction of one option is to say that he would take that option if forced to choose. It isn't the same as predicting he will take that option because you can describe him as leaning one way when you know that incoming information will make him go the other. So on this decision-theoretic account, we project our inclinations onto vague predicates creating an illusion of structure.

My reply to this deflationary counter-analysis begins with the question of why we lean. When nearly everyone leans the same way, there is normally a *basis* for their convergence. The counter-analysis cannot ground the widespread decision tendency in terms of the probable truth value of the borderline statement (or its perceived degree of truth). To avoid the implausible conclusion that this psychological uniformity is just a coincidence, the critic would have to say that we lean by virtue of our nature or nurture. But this is too sketchy. Until we get the prospect of details, we should defer to a corollary of the principle of charity: minimize the attribution of illusion.

2. *Concordance between Vague Terms*

The concordance between vague terms constitutes a second way that borderline regions are structured. We expect synonyms to share their borderline cases. If chopping off the head of two-headed man is borderline decapitation, then it is also borderline beheading. Concordance is also expected for symmetrically vague predicates. For example, 'noonish' means 'near noon', thus the truth of 'If 11:45 is noonish then 12:15 is noonish'. Semantically analogous terms should also sway together. If 'window' is precisified so that its hole reading prevails over its filler reading, then we also expect parallel precisification for 'door' and 'hatch'. This precedence effect can have philosophical implications. If we were to precisify 'person' to encompass corpses, we would also be inclined to count the "brain dead" as people. More interestingly, we would be more inclined to count fetuses as people. However, an anti-abortion position would not receive much support from this broad construal of 'person' because the dramatically lowered standards of treatment brought about by death would serve as an analogy justifying low standards of treatment for fetuses.

Concordance suggests that most vagueness theorists are too individualistic in their approach to borderline cases. Precisifying one term may require parallel precisifications of other terms in its wake. This holism has a second aspect: precisification of one term may require vaguification of another.

3. *Definers Vaguify as well as Precisify*

One of the main reasons clear cases are thought to have higher semantic standing than borderline cases is because we speak of precisification but not vaguification. This asymmetry suggests that definers are removing the inferior material and inserting better, more potent stuff.

However, etymology reveals both precisification and vaguification. Even if we restrict ourselves to the history of science, we find new terminology breeding new borderline cases. In addition to the unsystematic production of new vagueness, we find exploratory borderline regions. Consider the pattern by which medical predicates have eaten away at the extension of moral predicates. Instead of proceeding all at once, we find the borderline zone creeping deeper into the moral domain. The change is not so much a shrinkage of the zone as it is a shift: borderline medical issues become clear medical issues (leprosy, Tourette's syndrome, insanity) while clear moral issues become borderline ones (kleptomania, homosexuality, bastardry). A similar trend can be seen in the way our physical vocabulary has encroached on mental predicates. It is as if the re-zoning had a softening up function. Of course, the trends are not inevitable. Sometimes the borderline zone halts, sometimes it reverses or even follows a pendulum pattern (like our nature/nurture vocabulary). The borderline zone between the living and non-living wavers nowadays. Biochemists stress the arbitrariness of counting viruses as extremely simple organisms or as highly complex molecules. They are poised to move the borderline region in either direction. Successful extension of evolutionary concepts to borderline entities puts pressure behind their promotion to 'living things'. For instance, "natural selection" appears to have operated on the early atmosphere's catalysts when amino acids were first forming. Then again, subsumption of borderline entities under standard chemical explanations (such as of crystal formation) puts pressure in the opposite direction. The important thing to note is that the force shifting a borderline *F* into the clear *F* zone also tends to drag neighboring clear negatives into the borderline *F* zone. With new light comes new shadows.

This dynamic picture of borderline zones contrasts with the static picture presupposed by the supervaluationists. Their assumption of a predetermined twilight area seems an arbitrary departure from their theme of an evolving, opportunistic language.

Perhaps a liberalized descendent of supervaluationism could accommodate this dynamism. If we regard meaning as a good thing, then there is a presumption in favor of filling in the truth-value gaps so that the predicate becomes more meaningful. Although supervaluationists appear to operate under this assumption, one might suspect that this is too quick, an inference illegitimately influenced by an analogy with the underspecificity sense of 'vague'. Our attitude toward borderline cases could be as complicated as our attitude toward holes. Holes in boats and socks are bad but holes in Swiss cheese and doughnuts are okay. There are even some good holes; think of fish nets, hula hoops, and the holes in dashboards reserved for future accessories. Since holes can be helpful, we make new holes, turning the unholed into the holed. In addition to revealing a richer array of attitudes, the analogy with holes has the further advantage of letting our quasi-supervaluationist reflect the peculiar semantic status of borderline cases. He can say borderline cases are part of the meaning of a term in the degenerate way holes are part of the things containing them. In any case, if borderline cases also range from bad to good, supervaluationists should make room for vaguification.

Indeed, an early fellow traveller, John Kearn, did suggest that the liar sentence 'This proposition is not true' is incorrigibly vague.[40] His reasoning was that the proposition lacks a truth value and so since both ways of filling in the truth value lead to contradiction, it cannot be precisified. However, this is only an argument for leaving something vague, not for making it vaguer. Perhaps the contemporary prohibition against dividing by zero constitutes a full-fledged vaguification. Until the eighteenth century, thinkers were undecided as to whether the quotient was 1 or infinity. Nowadays expressions such as '4/0' are rejected on the grounds that they rest on a false presupposition (in this case, that there is a unique divisor n such that $0 \times n = 4$). Maybe some speculative interpretations of the wave/particle duality of light amount to a recommendation to make light indeterminate between 'wave' and 'particle'. Neutrality might offer the sort of advantages that enthusiasts of Spinoza claim for the "double-aspect" solution to the mind/body problem. Lastly, Heisenberg's uncertainty principle might be interpreted as a recommendation to vaguify previously precise predicates such as 'position' and 'velocity'.

In any case, the history of science contains many terminological revisions that resulted in new borderline cases. This occurs when absolute concepts are exchanged for ones that are sensitive to degrees. Adding argument places to an old predicate can raise unprecedented questions about what to relativize to. Vagueness also arises when a previously empty predicate is redefined to give it application: vacuum, split atom, invisible light, action at a distance. Borderline cases rush in along with the clear positives. A third way science generates novel vagueness is by conjoining operational definitions of asymmetric relations such as 'harder', 'louder', and 'brighter'. Each of the old definitions contained a test that was treated as a sufficient condition for the relation, so under the new complex operational definition, conflicting tests produce conceptual gridlock. Simplification of the term's instension can also produce conflict vagueness. When biologists defined 'female' as the sex that contributes more than half the reproductive material, they sacrificed a substantial part of the ordinary meaning in order to gain a short, quantitative rule of usage. Plants were previously thought to be devoid of females but the new definition changed about half of these clear negative cases to clear positives. The revision eliminated many borderline cases because its simplicity prevented clashes between rival criteria. Stereotypical female traits included: bearing and caring for offspring, smaller size, less aggression. Animals such as sea horses were puzzling because both sexes had female traits. The simple definition eliminated these borderline cases but it created others. Certain algae and protozoans produce gametes of uniform size. Since these isogamous organisms are sharing genes, their reproduction is sexual. But which sexes are involved? Since the gametes are uniform, they can be neither male, female, nor hermaphrodites. The borderline case did not arise for old concept of female because plants and minute organisms were considered sexless.

IX. CONCLUSION

When we are giving a purely reportive definition, we must preserve both clear cases and borderline cases because they contribute equally to the meaning of the term. When we engage in definition governed by desiderata that are not restricted to meaning preservation, *both* clear cases and borderline cases are revisable and in any direction (including from clear to borderline). Some of these desiderata will favor revisions that change borderline cases to clear cases. For instance, operational definitions put a premium on testability and so tend to turn most borderline cases into clear negatives. Psychologists who define 'hungry rat' as 'rat that has been deprived of food for 23 hours' also turn many clear positives into clear negatives. This is the typical pattern of operational definitions because they work by restricting the extension to a single (nearly) sufficient condition for applying the original term. Nevertheless, there is a difference of impact: borderline cases are driven out in favor of clear ones. Other desiderata such as completeness, consistency, problem-solving power, and simplicity will also appear to drive out borderline cases as long as we mistake the nature of vagueness or fall victim to the equivocations discussed in the second section. But once we command a clear view of vagueness, we are reconciled to our actual practice of preserving both the light and the shadows.

Department of Philosophy
New York University

NOTES

[1] D. M. Armstrong, 'Does Knowledge Entail Belief?' *Proceedings of the Aristotelian Society* vol. LXX 1969/70 (London: Methuen & Co., 1970), pp. 21—36; pp. 35—36. Another well-known advocate of postponing hard cases is John Rawls; see, for example, p. 96 of 'A Kantian Concept of Equality', *Cambridge Review* (February 1975). The methodological propriety of focusing on atypical and marginal cases is defended by Grover Maxwell and Herbert Feigl on p. 195 of 'Why Ordinary Language Needs Reforming' in *The Linguistic Turn*, ed. Richard Rorty (University of Chicago Press, 1967), and by Amartya Sen on p. 14 of 'Rights and Agency', *Philosophy and Public Affairs* vol. **11** no. 1 (Winter 1982).

[2] Actually, this ambiguity spawns a large family of verbal disputes and confusions. I discuss these in 'The Ambiguity of Vagueness and Precision', *Pacific Philosophical Quarterly* vol. **70** no. 2 (June 1989), pp. 174—183.

[3] Stanislav Andreski, *Social Sciences as Sorcery* (London: Ebenezer Baylis and Son, 1972), p. 70.
[4] Duhem makes this widely cited claim in *The Aim and Structure of Physical Theory*, trans. P. P. Wiener (Princeton, N.J.: Princeton University Press, 1954), pp. 178—179. Richard Robinson defends vague concepts with the alleged relation between vagueness and probability in *Definition* (Oxford: Clarendon Press, 1950), p. 184.
[5] I defend the epistemic theory in *Blindspots* (Oxford: Clarendon Press, 1988).
[6] Dreyfus's argument appears in *What Computers Can't Do* (New York: Harper, 1972).
[7] Jerry Fodor, *The Language of Thought* (Cambridge, Massachusetts: Harvard University Press, 1975), p. 63. Virtually all proponents of truth-conditional semantics subscribe to the doctrine Fodor is applying. For nice endorsements of the method see p. 92 of David Lewis's *Counterfactuals* and p. 36 of Mark Platts' *Ways of Meaning.*
[8] Ernest Nagel 'Some Reflections in the Use of Language in the Natural Sciences', *Journal of Philosophy* vol. **XLII** no. 23 (November 8, 1945), pp. 617—630; pp. 620—621.
[9] Raziel Abelson, 'Definition', *Encyclopedia of Philosophy* vol. 2, ed. Paul Edwards (New York: MacMillan, 1967), p. 315.
[10] See his 'Rejoinder: On a Kantian Conception of Language' in *The Paradox of the Liar*, ed. Robert L. Martin (New Haven: Yale University Press, 1970), pp. 59—66.
[11] Fine's classic treatment appears in 'Vagueness, Truth and Logic', *Synthese* vol. **30** nos. 3—4 (April—May, 1975), pp. 265—300.
[12] The quotation appears in Branden Matthews's *Parts of Speech* on p. 126.
[13] The term was introduced by Sidney Axinn and David Axinn in 'Notes on the Logic of the Ignorance Relations', *American Philosophical Quarterly* vol. **13** no. 2 (April 1976), pp. 135—143.
[14] Lycan applies this principle in 'Evidence One Does Not Possess', *Australasian Journal of Philosophy* vol. **55** (1977), pp. 114—126; p. 119. He also uses it in his books *Logical Form in Natural Language* and *Consciousness.*
[15] Derek Parfit, *Reasons and Persons* (Oxford: Oxford University Press, 1984), p. 213.
[16] W. V. Quine in a review of Milton Munitz's anthology *Identity and Individuation* in the *Journal of Philosophy* vol. **LXIX** no. 16 (September 7, 1972), pp. 488—497; pp. 489—490.
[17] Nelson Goodman, *The Structure of Appearance* (New York: Bobbs-Merrill, 1951), p. 6.
[18] Lawrence C. Becker, 'Human Being: The Boundaries of the Concept', in *Medicine and Moral Philosophy*, ed. Marshall Cohen et al. (Princeton: Princeton University Press, 1982), pp. 23—48; p. 47.
[19] Jonathan Glover, *Causing Death and Saving Lives* (Middlesex, 1977), p. 45.
[20] Michael B. Green and Daniel Wikler, 'Brain Death and Personal Identity' in *Medicine and Moral Philosophy*, pp. 49—77; p. 60.
[21] F. Nietzsche, *The Gay Science*, trans. Walter Kaufman (1974), p. 266.
[22] Johnston addresses the objection, that he is merely changing the subject in 'Relativism and the Self', *Relativism: Interpretation and Confronation*, ed. Michael Krausz (Notre Dame Press, 1989), pp. 441—472.
[23] J. Katz and J. A. Fodor, 'The Structure of Semantic Theory', *Language* vol. **39** (1963), pp. 170—210; p. 173.

[24] C. G. Hempel, 'Vagueness and Logic', *Philosophy of Science* vol. **6** (1939), p. 170. Max Black defended a similar theory in 'Vagueness: An Exercise in Logical Analysis', in his *Language and Philosophy* (Ithaca, New York: Cornell University Press, 1949).
[25] Ludwig Wittgenstein, *Philosophical Investigations*, p. 227.
[26] Daniel Dennett's evidence for the inconsistency of pain is given on pp. 221—225 of *Brainstorms* (Harvester Press, 1978).
[27] Ibid., p. 225.
[28] Ibid., p. 228.
[29] Ludwig Wittgenstein, *Zettel*, ed. G. E. M. Anscombe and G. H. von Wright, trans. G. E. M. Anscombe (Blackwell, 1967), sec. 403.
[30] Ibid., sec. 553.
[31] W. B. Gallie, 'Essentially contested concepts', *Proceedings of the Aristotelian Society* **56**, (1955—56), p. 172. For a recent discussion, see Peter Ingram's 'Open Concepts and Contested Concepts', *Philosophia* vol. **15** nos. 1—2 (September 1985) pp. 41—59.
[32] Norman Malcolm in *Consciousness and Causality* (co-authored with D. M. Armstrong) (Oxford: Basil Blackwell, 1984), p. 19.
[33] Norman Malcolm, p. 21.
[34] Quine's acceptance of partisan borderline cases can be gleaned from p. 41 of *Word and Object* (Cambridge, Mass.: MIT Press, 1960). For a more recent example, consider Dale Thorpe's characterization of borderline cases as ones in which opposite classifications are rationally acceptable on p. 391 of 'The Sorites Paradox', *Synthese* **61** (1984), pp. 391—421.
[35] Keith Lehrer, *Knowledge* (Oxford: Clarendon Press, 1974), pp. 61—62.
[36] Colin Radford, '"Analysing" Know(s) That', *Philosophical Quarterly* **20** (1970), pp. 222—229; pp. 228—229.
[37] Keith Lehrer, *Knowledge* (Oxford: Clarendon Press, 1974), pp. 61—62.
[38] Unger is the most prolific incoherentist. His most direct discussion is in 'Why there are no people', *Midwest Studies in Philosophy* (Minneapolis: University of Minnesota Press, 1981), pp. 177—222.
[39] Ludwig Wittgenstein, *Philosophical Remarks* (London: Basil Blackwell, 1975), p. 332. Many of the prophesized philosophers appear in the bibliography of Graham Priest's *In Contradiction* (Dordrecht: Martinus Nijhoff, 1987).
[40] Kearn says 'true' and 'false' are incorrigibly vague in 'Some Remarks Prompted by van Fraassen's Paper', *The Paradox of the Liar*, pp. 47—58; p. 54.

WILLIAM G. LYCAN

DEFINITION IN A QUINEAN WORLD

It has been nearly forty years since the publication of 'Two Dogmas of Empiricism.'[1] Despite some vigorous rebuttals during that interval,[2] Quine's rejection of analyticity still prevails — in that philosophers *en masse* have either joined Quine in repudiating the "analytic"/"synthetic" distinction or remained (however mutinously) silent and made no claims of analyticity.

This comprehensive capitulation is somewhat surprising, in light of the radical nature of Quine's views on linguistic meaning generally. In particular, I doubt that many philosophers accept his doctrine of the indeterminacy of translation, which directly underwrites the rejection of analyticity even though it did not figure prominently in 'Two Dogmas' but was made explicit only in *Word and Object* and subsequent papers.[3] (Indeterminacy of translation spells the death of individual word meaning as well as that of propositional or locutionary meaning, and so leaves no room for *truth by virtue of* word meaning as is required by analyticity.)

In this paper I shall make a Quinean case against analyticity, without relying on the indeterminacy doctrine. For I join the majority in denying both analyticity and indeterminacy: *Contra* Quine, I think there is determinate propositional meaning, as represented by a structured set of possible worlds, but with Quine I doubt that any sentence of a natural language is true solely in virtue of its meaning.[4] However, we shall see that the attack on analyticity is partly blunted by the actual practice of stipulative definition, and that practice will be my concluding focus.

1. QUINE AGAINST ANALYTICITY

As is well known, Quine begins by complaining that the "analytic"/"synthetic" distinction is imprecisely drawn. It should by now be equally well known that that is the least of Quine's complaints. He further objects to the circularities that infest philosophers' informal explanations of the "family circle" of intentional notions he is attacking:

J.H. Fetzer, D. Shatz and G. Schlesinger (eds.), *Definitions and Definability: Philosophical Perspectives*, 111–131.

'analyticity', 'synonymy', 'contradiction in terms', and so on. Grice and Strawson (*op. cit.*), among many other commentators, objected that *all* systems of definitions have this "circular" character. But Quine's point is that ordinary systems of definitions normally contain at least one term which is antecedently and nontechnically understood, in which case the circularity (of course) does not matter, while among the family-circle words there is no such term — they are all technical. (One might think that 'synonymy' is nontechnically clear enough, but as Gilbert Harman has emphasized, ordinary people use that word far more broadly and loosely than philosophers do.[5] One might also think that 'definition' is all right, but as we shall see, 'definition' is a morass.)

Yet the real objection to analyticity is neither the imprecision nor the unspecifiability of a distinction between analytic and synthetic; the first half of 'Two Dogmas' was historically very misleading in this regard.[6] *Defining* analyticity is not the problem. In fact, analyticity is rather easily defined, as a strongly modalized variety of truth: not just nomologically or even metaphysically necessary truth, but "conceptual" truth, truth by virtue of lexical meaning alone. A sentence is analytic iff its own meaning suffices to make it true, regardless of any (other) contribution from the world.

This way of understanding analyticity has obvious epistemological consequences: An analytic sentence would be *unrevisable*, in the sense that to deny or reject it would be *eo ipso* to abandon its standard meaning; one who called it false would be, as Quine says, not denying the doctrine but changing the subject. Thus nothing could count as evidence against the truth expressed by an analytic sentence, and more generally we could have no rational grounds for doubting that truth (we could be mistaken only about the meanings of the relevant words).

Quine's real complaint is that analyticity thus understood is just unexemplified. There are no analytic sentences, because no truth is necessary in any stronger sense than the nomological, and no belief is unrevisable. (There is, if you like, "analyticity" relative to a theory on an axiomatization of that theory: A sentence may be *treated as* analytic within a theory, by being used as an axiom and introduced in textbooks by an axiomatic definition. But mere analyticity-within-a-theory-on-an-axiomatization is not what partisans of analyticity have had in mind.)

So let us review some attempts to explain how a sentence can be both necessary in some stronger sense than that of physical or causal necessity ("logically," "conceptually," etc.), and epistemically unrevisable in the way described. There are three main kinds of account expressed or presupposed in the literature. The first two I shall mention but dismiss briskly, in part because they are unpromising and in part because they have little to do with the topic of this volume. It is the third that will detain us.

II. THE FIRST UNPROMISING ACCOUNT

The first account is a straight shot: the idea that truth in virtue of meaning is just that. Sentences have meanings; some meanings are such as to guarantee truth, and that is (virtually) all there is to it. Call this the "Simple Account."

At the time of 'Two Dogmas' *meanings* were thought of as propositions, language-neutral abstract entities "expressed" by sentences. The positing of propositions was supposed to explain the properties of language-bound sentences individually and in pairs — meaningfulness, ambiguity, synonymy, entailment, and the like — in terms of the relation of expressing. Psychological facts about the propositional attitudes were complementarily to be explained with the aid of the further relation of "grasping" between a subject and a proposition.

Understood in these terms, the Simple Account of analyticity has it that a sentence is analytic just in virtue of expressing a proposition of a special sort — the null proposition that rules out no contingent possibility. Alternatively, one might say that although an analytic sentence is well-formed and trivially or degenerately true, it expresses no genuine proposition. Either way, such a sentence — meaning what it does mean — cannot be false.

Quine's objection to this propositional theory of analyticity is that it invokes propositions. His main mature objection to positing propositions is that if there really were propositions then translation would be determinate: yet translation is indeterminate. However, I myself am sworn not to rely on the indeterminacy doctrine, and so must find other objections. Quine himself makes two, and a third can be pressed on his behalf.

The first objection is akin to the indeterminacy claim, but does not rely on it: The proposition theory does not square with the translation habits of ordinary people and with their ordinary talk of synonymy and the rest. In the real world, people count as "synonymous" any two sentences that serve roughly the same contextual purpose on an occasion, and such pairs will seldom be candidates for doing what philosophers call expressing the same proposition. Ordinary people rather engage in the loose free-for-all practice that Quine calls "paraphrase" (Sec. 33 of *W&O*, p. 208, and Secs. 53—54, and see Harman, p. 142). Secondly, propositions themselves are *entia non grata*, "disorderly elements," "creatures of darkness," etc.; they are nonspatiotemporal and acausal, their identity conditions are inscrutable, and so forth.[7]

The third objection is emphasized on Quine's behalf by Harman (*op. cit.*). It is that propositions are *flatus vocis* or dormitive virtues, and do no explanatory work whatever. ("Why are these two sentences synonymous?" — "Because they express the same proposition." — "Oh, I see.") The proposition "theory" of meaning is a pseudo-theory, not a substantive theory at all.

I disagree with Harman on this general point, for I think the theory of propositions is a genuine and even substantive theory, even though a very bad one.[8] But there is a serious question of the explanatoriness of the proposition theory with respect to analyticity in particular. A sentence is supposed to be true, and true in virtue of meaning, because it "expresses the Null Proposition." Like Harman, I fail to discern the explanatory force of that 'because'. Even if we know what "expressing" is, what do we know of the "Null Proposition" except that sentences that express it are supposed to be analytically or conceptually-necessarily true?

One may of course attempt to answer that rhetorical question nonrhetorically, by offering a story about what "propositions" are that says in particular what makes the "Null Proposition" special. The possible-worlds story comes instantly to mind, for propositions can be taken to be sets of worlds. Indeed, David Lewis' theory of analyticity is an instance of the Simple Account by way of propositions by way of possible worlds: For Lewis, a sentence is analytic iff it is true at every world — and so expresses, though Lewis does not use the term, the Null Proposition, i.e., the proposition that excludes nothing.[9] Thus Lewis avoids my foregoing criticism of the Simple Account, and advances the issue by explicating analyticity as truth in all possible worlds. An analytic sentence is true in virtue of expressing the proposition that it does, the Null Proposition, for that proposition is the set of *all* worlds and is nowhere false.

Lewis maintains that the truth of a proposition throughout all worlds is a mind-independent fact of the pluriverse. Thus for him an analytic statement is true in virtue of the modal structure of the world, in virtue of what is objectively possible. In a sense, then, although a Lewis-analytic sentence is true in virtue of its meaning, it is not true *solely* in virtue of that but partly in virtue of an objective feature of the pluriverse. That is perhaps only a quibble, but in any case, there ensues the more trenchant Quinean question: Why should anyone think there are any sentences that are true in all possible worlds, if "possible" means something seriously weaker than nomologically possible? Lewis' theory makes no headway at all against this most basic misgiving.[10]

Might one perhaps pursue the Simple Account *without* reifying meanings, i.e., without appeal to propositions at all? That would be to hold that the truth of certain "necessary" sentences is still explained *solely* by reference to considerations of meaning, though now in some ontologically non-toxic sense. But if the Simple Account is to be left at that, we are given no hint of how the explanation of truth by meaning alone could proceed. We are told "how" sentences can be true by virtue of their meanings alone, only by being told that those special sentences' meanings suffice to make them true. Unless some entities comprising a mechanism are posited — at least propositions and/or possible worlds — no explanation emerges at all.

III. THE SECOND UNPROMISING ACCOUNT

The second unpromising account of analyticity — which historically replaced the Simple Account so far as the latter modestly appeared in Russell and Moore — is the positivist/verificationist theory of truth in virtue of *unfalsifiability*. A sentence that is verified by every possible experience, or at least falsified by no possible experience, would be a candidate for analyticity. And this Verificationist Account avoids Quine's and Harman's objections to the Simple Account, for it is genuinely substantive and explanatory. Verificationism in general is an undeniably explanatory theory of meaning, even if a false one, and it entails in particular that if a sentence is radically unfalsifiable, there is something very special about that sentence's meaning: so far as the sentence is true, it is true solely in virtue of meaning (for its meaning, i.e., its verification-condition, affords no room for falsity).

There are basically three objections to the Verificationist Account of analyticity. First, how could we ever know that a sentence had this character of in-principle unfalsifiability? At least, how could we possibly know that a sentence was immune in this way to refutation, *unless* we had some deeper antecedent reason to think the sentence was impregnable, e.g., that it was already true by convention. A look at the more accessible Positivist writings suggests that irrefutability is not itself the *explanation* of a truth's "conceptual necessity"; rather it is explained *by* necessity, which is in turn explained by something more familiar.

It is sometimes urged against such points that we know a sentence to be unfalsifiable simply because we cannot imagine or conceive the sentence's denial. Quine's answer to someone who pleads just plain inability to imagine a thesis' being false is simply that the person has a poor imagination. What one can imagine at any given time is almost entirely dependent on one's current theory, as the history of science and the history of ideas indicate; many great intellectuals of the past were "unable to imagine" many of the things which today are commonplaces.[11]

In any case (second objection), the operative term in the slogans "verified by all possible experiences," "falsified by no possible experiences," and the like is, of course, '*possible*'. The Verificationist Account cannot do without this qualification, and "possible" is a member of the family circle. Explaining necessary truth in terms of "possible experiences" is much like explaining it in terms of "possible worlds." Once again we must ask why certain imagined experiences — e.g., as of seeing an object that is simultaneously red all over and green all over, or as of a square circle — are supposed to be impossible in any stronger sense than that of nomological impossibility. To insist that such experiences just are inconceivable in that very strong sense adds nothing to one's initial insistence that the corresponding negative sentences are analytic.

Quine's most famous objection to the Verificationist Account (and my third) is a corollary of his Duhemian point against atomistic verificationism generally: that "statements about the external world face the tribunal of sense experience not individually but as a corporate body" ('Two Dogmas,' p. 41). Any single sentence of our total theory may be retained in the face of apparently recalcitrant experience, if enough more-or-less drastic revisions are made elsewhere in the theory. Thus no individual sentence *has* its own specific verification-condition. Someone might reply that while Duhem's point is true of contingent and empirical sentences, it obviously fails for sentences which are true or false no matter what; but that would, as before, beg the question of whether there are any such sentences in the first place.

Happily (for them), the Logical Positivists had a positive account of analyticity, in addition to experiential unfalsifiability, their negative criterion. It was that analytic sentences are true *by convention*. The Convention Account of analyticity is the third and last view that I shall consider. Notice that it is entirely independent of the Verificationist Account, even though the Positivists held both; historically, it was preserved among the Ordinary Language philosophers who rejected verificationism, and continues to be accepted by at least one (otherwise) hard-core scientific realist.[12]

A self-respecting contemporary metaphysical realist and referential semanticist finds the idea of "truth by convention" grotesque. How could a sentence be *true*, i.e., correspond to reality in a referential manner, entirely *by convention*? But of course the Positivists and the Ordinary Language philosophers were not (thoroughgoingly) either metaphysical realists or referential semanticists. More to the point, the idea of truth by convention is defensible and the Convention Account of analyticity is far more compelling than was either of its two predecessors.

IV. TRUTH BY CONVENTION

There are three main ideas of truth by convention. One is that of lexical meaning as codified in authoritative dictionaries of natural languages. The second is that of lexical rules or "meaning postulates" codified in formal languages. The third is that of truth by *stipulative* definition.

The dictionary idea is that word meanings are conventional and that the governing conventions are codified by the entries in a good dictionary for the language in question. Any "information" contained in or easily read off from a proper dictionary is thus purely conventional, and cannot be doubted or questioned by anyone who has mastered the language.

An obvious objection to this idea is that, in the real world, dictionaries are not stone tablets bearing our conventions in axiomatic form. *Most* entries in a typical dictionary do not even purport to codify analytic truths, and none does anything whatever to distinguish mere factual information from what philosophers would consider strictly parts of the meanings of the *definienda*; usually the entries contain just informal explanations of how to apply the concept or recognize the item in question, recognition criteria mixed with a motley of commonsensical facts. Hence Quine's apt remark that dictionaries differ in no principled way from encyclopedias.

One might respond that in real life, lexicographers get sloppy (many of them are free-lancing, and all are underpaid); that is just real life for you. But Quine has a more powerful objection to the dictionary idea: Considering the suggestion that "'bachelor' . . . is *defined* as 'unmarried man'," Quine asks, "Who defined it thus, and when?" ('Two Dogmas,' p. 24). This question is sometimes passed by as a rhetorical flourish or as merely a superficial piece of sarcasm, but it has two deep points. One is that on pain of regress, conventions of language could not all have been established explicitly by syndics sitting in council. (There are infinitely many logical truths. But there could not be a complete infinite totality of single conventions which fix all the logical truths, one by one, immediately. Therefore if logical truths are true by convention, they must be given by general conventions, we must rely on an antecedent knowledge of the very same sort of logical truths, in mediating our inferences. Thus the regress.) Quine originally took this to embarrass the idea of a linguistic convention quite badly, but he agrees it has been circumvented by Lewis' (*op. cit.*) account of tacit convention.[13]

The second and more telling point is that the lexicographer is in exactly the position of the field linguist in a case of radical translation; the lexicographer merely observes language already in use and theorizes about it. Thus he or she cannot experimentally distinguish between sentences that are fixed by convention as "analytic" in the target language and those that are merely considered *obvious* by the native speakers, since all that meets the eye and ear is one class of "stimulus-analytic" strings, the sentences that command universal assent and whose hypothetical denials elicit funny looks.

Granted, a lexicographer's evidence base is not exhausted by the linguistic behavior of others. The lexicographer who is a native speaker may appeal to his or her own linguistic competence and internal sense of word meanings. But the same criticism applies to that appeal, as it does to the Ordinary Language philosopher's introspective test of meaning ("Would we still call a thing an *X* if it weren't a *Y* any more?"): We cannot say even of ourselves, simply by introspecting, whether some belief of ours is true by definition or whether it is merely a very well-entrenched truism. For we have nothing to go on but the feeling of obviousness and thorough-entrenchedness. One might say that if the lexicographer knows what genuine synonymy is, he or she is entitled to use that notion in constructing an ideal, nonsloppy dictionary of a language, and can make the appropriate distinctions on sight. But this notion of "genuine synonymy" is just what is in question, and claimed by Quine to be an unexemplified philosophers' artifact. The appeal to existing or even idealized real-world dictionaries cannot discredit Quine's claim.

Perhaps ordinary natural language is not the proper locus of analyticity. Formal languages are better suited to foment analyticities, since they are under the control of logicians and philosophers in a way that real natural languages are not. And Carnapian "semantical rules" or "meaning postulates" are splendid candidates for analyticity. They are assimilated to the *definitional postulates* of an axiomatized theory.[14] A logical consequence of a definitional postulate, or at least the postulate itself, ought to be true in virtue of meaning alone.

Quine has a good deal to say about this;[15] two criticisms emerge. First, postulates are postulates only relative to a given context of inquiry; they can be changed around at will without this effecting any substantive change in the theory they inhabit. Notoriously (N. R. Hanson's example[16]), the second law of classical mechanics, $F = m(d^2s/dt^2)$, can be taken or employed variously in the organization of theory — as an axiom, as a stipulative definition of F, as a definition of m, or as a purely empirical generalization, depending merely on how the containing theory is axiomatized.

A Positivist might well reply here that postulates can indeed be added and dropped at will for reasons of convenience, and theories re-axiomatized, but that such tampering necessitates corresponding changes in the meanings of the terms which occur crucially in (and are "implicitly defined by") the theories; different axiomatizations give different meanings to the theoretical terms, even though the theories do not vary in overall empirical content. But Quine rejoins that there is no independent evidence at all of any such meaning changes having occurred, when the postulates of actual theories have been readjusted in real life. Put the matter up to any scientist or any intelligent but philosophically untainted specialist in some other cognitive discipline, and neither informant will know what we are talking about; though they might grant loosely that "what we understand by" a theoretical term has changed when we reaxiomatize or introduce a new definitional postulate, they would never maintain that the term had "just plain changed its meaning," or that it now simply meant something different from what it meant before. (Hilary Putnam gives numerous examples of relevant scientific changes that plainly have involved no meaning changes in any but a question-begging sense.[17]) The Positivist insists that the changes occur, because they must occur if the Positivist theory is correct. But on pain of severe *ad hoc*ness, the meaning changes would have to be exhibited in a non-question-begging way.[18]

Quine's second criticism of the appeal to definitional postulates is that theories can turn out to be false; so "truth by convention" does not guarantee *truth*. (Harman puts the criticism in this way on pp. 130–131.) As always, someone may reply that a theory's definitional postulates are special, in that even when their containing theory goes down in flames, their analyticity exempts them from falsification. But as always, the alleged analytic exemption is just what is at issue.

One might think that at least the theorems of classical logic would be exempt, before any special vocabulary had been introduced. And the notion that they hold purely in virtue of the meanings of the logical constants is a very appealing philosophy of logic, whether or not one approves of "truth by definition" in science or in other empirical discourse about the material world. But the present point holds for logical truths also: As the Law of Excluded Middle is questioned both (for epistemological reasons) by intuitionists and (for scientific reasons) by quantum logicians, so *might* we find epistemological or scientific reasons for qualifying even the Law of Noncontradiction. Not even logic is epistemically unrevisable in the strong sense required for analyticity. Moreover and contributorily, the alleged distinction of kind within linguistic semantics between the "logical constants" and other words of English is dissolved by even a little examination.[19]

Let us turn to the third idea of truth by convention, inspired by the occasional practice of *novel* stipulative definition within more or less natural languages. Suppose I offer a novel coinage:

Dudent $=_{\text{def}}$ Dumb student.

And suppose I add, now employing my newly augmented object language, "All dudents are students, you see." How could I possibly be mistaken? My assertion seems a clear example of the truly unrevisable. To question it would be to question either the success or the wisdom of my stipulation, and to reject it would be to reject the stipulation and refuse to join in speaking the augmented language; my audience could not "deny the doctrine" and genuinely disagree with what I had said, for the ostensible denial, "There are dudents who are not students," simply makes no sense in the context. It is either a straightforward self-contradiction or it means nothing at all.

Quine himself is startlingly respectful of this idea:

> There does . . . remain still an extreme sort of definition which does not hark back to prior synonymies at all: namely, the explicitly conventional introduction of novel notations for purposes of sheer abbreviation. Here the definiendum becomes synonymous with the definiens simply because it has been created expressly for the purpose of being synonymous with the definiens. Here we have a really transparent case of synonymy created by definition; would that all species of synonymy were as intelligible. For the rest, definition rests on synonymy rather than explaining it. ('Two Dogmas,' pp. 25–26)

This apparent concession is larger and more damaging than perhaps Quine realized in making it. For it seems to allow that there are analyticity-generating synonymies and so genuine instances of full-fledged analyticity, even if they are few. Moreover and worse, if the introduction, for notational convenience, of new abbreviations for familiar but cumbersome terms can yield analytic truths, then the way is opened to a whole further range of analyticities. For if we are ever able to make two sentences equivalent in meaning by fiat, then presumably it is also possible for us to find out conclusively, to hypothesize, or even to "reconstruct" *a la* Hobbes and the State of Nature, that someone already has made other pairs of sentences equivalent by fiat in the past. So, it seems to me, a good Quinean ought to argue as powerfully as possible against the possibility of creating genuine analyticities even by explicit stipulation. Accordingly I shall take up and try to expand some of Quine's and Harman's objections to the Stipulative Definition account of analyticity, adding two of my own as well.

IV. STIPULATIVE DEFINITION

The obvious Quinean objection is based on the indeterminacy doctrine: Suppose a friend comes by and offers a stipulative definition, say: "'Veline' shall mean *vegetarian cat*." We could accept a translation manual according to which our friend's phrase 'shall mean' is taken onto the term 'obviously denotes' and suitable readjustments of syntax are made. The idea is that our friend, in her role as unspoiled native, behaves toward the indicative in question, “All velines are vegetarian cats,” exactly the same as she behaves toward an admittedly synthetic but trivial and obvious truth — except for the attachment of a particular label: “analytic,” “$=_{def}$,” “shall mean,” or what have you. And our translation of the *label* is indeterminate; there is no fact of the matter. And so there is no fact of the matter as to whether our friend's utterance is really a stipulative definition or merely the reiteration of an obvious truth for the common good.

On Quine's view, her utterance is indeterminate in a second way as well: There is no fact of the matter as to what her putative *definiens* means. Even if we were able to bypass the indeterminacy of “$=_{def}$” & Co., and were able to say determinately that “veline” was analytically connected to another expression, we would not thereby know a necessary truth unless we knew what the *definiens meant*, to the exclusion of what it coextended with but did not mean — and this, according to the indeterminacy doctrine, we cannot know, because there is no fact of the matter to be right or wrong about.

As before, I shall not rely on the inderminacy doctrine. But the first of these two appeals to translation has some force independently of the indeterminacy claim.

Objection 1

On p. 27 of ‘Two Dogmas,’ Quine lists three possibilities as to what is going on when new notation is stipulatively defined in a formal system. Of the first two (both technical equatings of long technical locutions with shorter terms, but for different purposes) he complains, as usual, that they already presuppose a prior notion of strict, analyticity-generating synonymy and so cannot be appealed to in an argument for the existence of analytic sentences.[20] In keeping with his concession quoted above, he expressly refrains from bringing the same complaint against the third possibility that “. . . the definiendum may be a newly created notation, newly endowed with meaning here and now.”

But we may bring it. The idea of adventitious stipulative definition *as it would have to be construed for the purpose of saving analyticity* must presuppose the notion of strict, analyticity-generating synonymy. Otherwise it would collapse back into mere definitional postulation. Stipulators must have the power to create synonymies and truths by fiat. The “dudent” example seems to show that people do sometimes have that power. But notice that the "dudent" argument still presupposes the analyticity of logical truths: "All dudents are students" is true because 'dudent' has "student" as part of its meaning and all *students* are students (cf. 'Truth by Convention,' pp. 71—72). As we have seen, Quine has argued against the analyticity (in the strong sense) of the logical truths; so he would hardly give in on that of sentences obtained from logical truths by substitution. Recall Harman's point that a sentence's being "endowed with meaning" on the spot does not prevent it from being false and thus does not *solely* explain its being true.

Notice too that if "All dudents are students" is truly analytic, its syntactic denial is meaningless; if a sentence really is true *just* in virtue of its meaning, then that sentence's syntactic denial cannot mean what it appears to say, and so either means something else or means nothing. But "There are dudents who are not students," though (if you like) logically false, is not meaningless. We all know what it means; if we did not, we would not know that it was logically false and that to token it is to contradict oneself.

Objection 2

Let us revisit and generalize the first of Quine's two arguments against analyticity through definitional postulation (cf. Harman, p. 141). The main problem was that the same theory or belief system can be axiomatized in different ways, treating different elements as "analytic." More formally:

Suppose

(1) A sentence S in theory T_1 is "true by definition."

And

(2) The very same sentence S appears in theory T_2 as a purely empirical generalization.

(Let S, for example, be "$F = m(d^2s/dt^2)$.")

(3) T_1 and T_2 are merely two different axiomatizations of more or less the same body of truths.

(4) A sentence's analyticity/syntheticity is a function of the meanings of its component terms. [Tenet of the "analyticity" view.]

(5) Analytic sentences are those which are "true by definition"; others are synthetic. [Assumption for *reductio*.]

But

(6) The meanings of S's component terms do not change (in any ordinary sense) from T_1 to T_2. [Seemingly supported by (3), but also plausible on its own.]

(7) Either S is analytic in both T_1 and T_2 or S is synthetic in both T_1 and T_2. [(4), (6)]

(8) S is analytic in T_1 but synthetic in T_2. [(1), (2), (5)]

(9) Contradiction! [(7), (8)]

(10) (5) is false. [Since our premises (1), (2), (3), and (6) are true and we grant the analyticity theorist (4).]

So "definition does not hold the key to analyticity."

This formulation of the Quine-Harman argument opens the way for the standard reply simply put: that premise (6) begs the question, because on the "analyticity" view it follows trivially from (1) and (2) that the meanings of the component terms of S do change.

Here as in the previous, more specific case of definitional postulation, I am in sympathy with Quine's and Harman's replies on this point: Who says the meanings change from T_1 to T_2? It *seems* that all we have here is two organizations of the same theory differing only in elegance or in convenience for particular purposes. It is up to the defender of analyticity to show why that natural reaction is wrong. But the analyticity theorist cannot do this without relying on the essential tenets of just the philosophical view that is in question. Thus, it is the analyicity theorist, not Quine and Harman, who is begging the question here if anyone is.

Just as there is no difference between a "meaning postulate" and an ordinary postulate (in the sense of a very general truth chosen to generate other truths), that has other than *ad hoc* considerations to recommend it, there is no non-*ad hoc* difference between a sentence "true by stipulative definition" and one simply true *ex hypothesi*. Even stipulative definitions, like postulates, can rationally be given up in physics without apparent change of meaning. Although the clearest cases of postulational change without meaning change involve natural-kind terms when new empirical information comes in, there are other, more ordinary occasions for semantically harmless readjustment of taxonomies as well.[21]

Objection 3

In stipulative definition, there is an unexpected problem of *getting from* the stipulator's token to an indicative sentence held to be analytically true. The problem takes a bit of explaining.

One strong temptation to see stipulative definition as yielding (however harmless) analytic truths arises from the fact that, when a person utters (tokens) an explicit stipulative definition, one cannot at all appropriately respond, "That's false." (Lucy: "Here's my first definition: A 'freebish' is a dog eating pizza." Linus: "That's false, Lucy, because . . . " Sound effect: *POW*!) Naturally enough, this is felt to show that explicit stipulative definitions and their consequences *cannot be* false, and thus are necessarily true.

I shall show here (expanding on pp. 71—72 of 'Truth By Convention,' *loc. cit.*) why this approach to necessary truth is blocked. Let us first look at some typical tokens of explicit stipulative definition:

(a) 'Veline' $=_{\text{def}}$ vegetarian cat.

(b) All velines are by definition vegetarian cats.

(c) We shall use 'veline' to mean "vegetarian cat."

(d) Let 'veline' abbreviate 'vegetarian cat'.

(e) 'Veline' shall denote vegetarian cats.

(f) Let 'veline' mean *vegetarian cat.*

Now none of these tokens is *itself* a straightforward indicative sentence alleged to be analytic. It is the prior occurrence of one of these that it supposed to legitimize the imputation of analyticity to some later, distinct sentence: "All velines are vegetarian cats." So we may well ask how this process takes place, and what sort of relation holds between the tokens (let us hereafter call them "stipulatives") and the indicative, "All velines are vegetarian cats."

Stipulatives (d) and (f) best capture the spirit of convention-fixing, being hortatory subjunctives. Obviously they cannot be true or false;

they state *rules*, more or less in the form of *commands* from the Great Convention-Giver to her people. (c) and (e), I think, are best interpreted as announcements of the speaker's intention to obey a certain rewriting rule — the idea being that if you want to understand the speaker, you must obey the same rule *pro tem*. (c), then, is just a slightly more tactful version of (d).

It is hard to assess (b)'s logical status; (b) seems to be an indicative, not a subjunctive or a command. But one could, I suppose, state the stipulative definitions of one's theory in this form, so long as one used a heading on the page which made it clear that the sentences codified stipulative and not reportive definitions. (a) could be read aloud as almost any of the other stipulatives listed; it does not seem to have a separate logical status of its own.

So we have two basic kinds of stipulative: A hortatory subjunctive, or perhaps an imperative telling us to do such-and-such under such-and-such conditions; and an indicative sentence that either is flagged (by a key phrase such as "by definition" appearing within it) or is prefaced by a heading such as "Abbreviations" or "Defined Terms" or even "Semantical Rules." In what way do stipulatives of either kind yield necessary truths?

Let us take the latter kind first. (b) does seem to entail "All velines are vegetarian cats," but what entitles us to conclude that the latter sentence is necessary in the sense of analyticity? Only the presence of the flag, the phrase "by definition," or perhaps the appearance of (b) under a heading of the kind just mentioned. And it is the forces of just such phrases and headings that is in question here, the question of whether a speaker can make a sentence true by announcing that it shall be so. So the stipulative occurrence of (b) does not suffice to distinguish (b) from the other stipulatives or to show that "All velines are vegetarian cats" could not possibly be false. So let us go on to the more promising possibility of getting analytic truths directly from (rewriting) *rules* of the language itself.

How is "All velines are vegetarian cats" obtained from, say, (d), or from a still more explicit rule, "Rewrite 'vegetarian cat' as 'veline'"? There is no entailment here (even if we could make sense of the notion of entailment independently of all the talk about meaning and necessary truth); since (d) is not an indicative, it has in the ordinary sense no truth value and hence cannot entail anything. The only obvious way to get "All velines are vegetarian cats" from our rewriting rule is to regard it as being the result of applying the rule to the logical truth, "All vegetarian cats are vegetarian cats." But there is still no reason to regard "All velines are vegetarian cats" as being true by convention, even though it is in part the product of a rewriting rule or conventional abbreviation which *is* (we may concede) in a clear sense purely a linguistic convention; because it has not been shown that the logical truth originally operated on by the abbreviative rule is true by convention.[22] We could regard "All velines are industrious" as being the result of applying our rewriting rule to "All vegetarian cats are industrious," but that does not make the former a necessary truth.

A second way of putting Objection 3 is to point out that any way of getting a truth couched in the object language from an explicit or tacit stipulative must make a significant use/mention move and that that move is sure to be fallacious. Stipulatives are metalinguistic, besides being hortatory or proclamatory in mood, while (in terms of truth conditions) the corresponding object-language trivialities are directly about the world, however little they illuminate it. Therefore the object-language truisms do not follow deductively from the stipulatives; and there is no *other* reason to think that the stipulatives make them analytic. (There is no objection to call them "definitional truths," so long as it is understood that 'definitional' does not mean analytic. Nor need we balk at calling *expressly* logical truths logical, any more than Quine would object to calling laws of physics nomological truths, or to calling laws of the state of Massachusetts legal truths. The theorems of a particular system of logic form a well-defined and interesting class. It is just that they are not true (solely) by definition or by virtue of meaning.)

Incidentally, we now have a way of blocking the argument from the inappropriateness of "That's false," mentioned above: The reason "That's false," uttered in response to a stipulative definition, is inappropriate or sounds funny or whatever is not that stipulative definitions cannot-be-false in the sense of analyticity. They are barred from falsity in the trivial sense of being truth-valueless by virtue of their syntax. In the same sense they cannot-be-true.

I close with an objection to "truth by convention" generally: It is only recently that anyone — Lewis[23] — has made clear sense of the notion of a "convention" in the first place. Lacking such an analysis, Quine's positivist and Wittgensteinian opponents had no direct way of testing their intuitive idea that certain special sentences were true "by convention." But using Lewis' analysis, we can test it. We can plug the idea of *holding something true* or *treating something as true* into Lewis' analysis and see if the result is plausible. (Lewis' analysis is surely accurate enough to use in such a test, though questions have been raised about its details.[24])

Lewis' formula is as follows (p. 78):

A regularity *R* in the behavior of members of a population *P* when they are agents in a recurring situation *S* is a *convention* if and only if it is true that, and it is common knowledge in *P* that, in almost every instance of *S* among members of *P*,
(1) almost everyone conforms to *R*;
(2) almost everyone expects almost everyone else to conform to *R*;
(3) almost everyone has approximately the same preferences regarding all possible combinations of actions;
(4) almost everyone prefers that any one more conform to *R*, on condition that almost everyone conform to *R* ['any one more' is a weakening of the stricter 'everyone', still in the spirit of "the more the better"];
(5) almost everyone would prefer that any one more conform to *R′*, on condition that almost everyone conform to *R′*,
where *R′* is some possible regularity in the behavior of members of *P* in *S*, such that almost no one in almost any instance of *S* among members of *P* could conform both to *R′* and to *R*.

Since Lewis has already co-opted the variable *S*, let us use *T* to stand for our putatively analytic sentence. For the case of "treating *T* as true," the "recurring situation *S*" is either vacuous or a (rare) situation in which one is called upon to assent to *T*. Now, according to Lewis' analysis, it is a *convention* among members of a population *P* to treat *T* as true iff it is true that, and it is common knowledge in *P* that, in almost every instance of *S* among members of *P*: (1) almost everyone treats *T* as true; (2) almost everyone expects almost everyone else to treat *T* as true; (3) almost everyone has approximately the same preferences regarding all possible combinations of actions; (4) almost everyone prefers that any one more treat *T* as true, on condition that almost everyone treat *T* as true; (5) almost everyone would prefer that any one more treat *T* otherwise, on condition that almost everyone treat *T* otherwise, where 'otherwise' indicates the possible but unchosen alternative to the actual behavioral regularity that is required for the regularity's being genuinely conventional.

The foregoing statement does not make a great deal of sense. (1) is fine, and (2), (3) and (4) are all right, though somewhat emptily so; but (5) rings true only if understood as supposing that the sentence *T* might mean something other than what it does mean. We already know that the truth of *T* is conventional in *that* way. What makes no clear sense is that, given *T*'s meaning what it means, we "would prefer that any one more treat *T* otherwise [than true], on condition that almost everyone treat *T* otherwise." Since *T*, meaning what it means, logically (not: analytically) cannot be otherwise than true and everyone knows that, we do not know what we should prefer if almost everyone were to start treating *T* otherwise than as true. Treating *T* as true is not like driving on the right-hand rather than the left-hand side of the road. Thus a friend of truth-by-convention has a heretofore unanticipated task: to show how Lewis' analysis of convention or some suitable successor applies to the truth of a sentence, in order to make it plausible that that sentence really *is* true *by convention.*

I think the arguments suffice to show that there is not much hope for an account of analytic, "conceptual," or supernaturally necessary truth based on stipulative definition.[25] And therefore (if I am right in thinking that the Stipulative Definition account offers our *best* hope) there is not much hope for account for such truth at all. Moreover we have reached that conclusion without joining Quine in his radical *general* nihilism about linguistic meaning. The onus is on the defender of analyticity actually to do some defending — to make a wholly new case for analyticity before relying any further on such notions.

Department of Philosophy
University of North Carolina

NOTES

[1] W. V. Quine, 'Two Dogmas of Empiricism,' *Philosophical Review* **60** (1951), 20—43, hereafter just 'Two Dogmas.' Reprinted in *From a Logical Point of View*, Second Edition (New York: Harper Torchbooks, 1961); all page references are to the latter.

[2] Gustav Bergmann, 'Two Cornerstones of Empiricism,' *Synthese* **8** (1953), 435—452; H. P. Grice and P. F. Strawson, 'In Defense of a Dogma,' *Philosophical Review* **65** (1956), 141—158; Jonathan Bennett, 'Analytic-Synthetic,' *Proceedings of the Aristotelian Society* **59** (1959), 163—188; Jerrold J. Katz, 'Some Remarks on Quine on Analyticity,' *Journal of Philosophy* **64** (1967), 36—52, 'Where Things Now Stand with the Analytic-Synthetic Distinction,' *Synthese* **28** (1974) 283—319, and elsewhere; and of course many others.

[3] *Word and Object* (Cambridge, MA: MIT Press, 1960), hereafter *W&O*; 'Ontological Relativity,' *Journal of Philosophy* **65** (1968), 185—212; 'Reply to Chomsky' and 'Reply to Hintikka,' in D. Davidson and J. Hintikka (eds.), *Words and Objections: Essays on the Work of W.V. Quine* (Dordrecht: D. Reidel, 1969); 'On the Reasons for Indeter-

minacy of Translation,' *Journal of Philosophy* **67** (1970), 178—183; 'Indeterminacy of Translation Again,' *Journal of Philosophy* **84** (1987), 5—10.

[4] Howard Callaway has also championed meaning without analyticity; see his 'Semantic Theory and Language: A Perspective,' *Philosophical Topics* Supplementary Volume (1980), 61—70, and 'Meaning without Analyticity,' *Logique et Analyse* **28** (1984), 41—60. Cf. also sec. 2 of the Introduction to Georges Rey's *Mind without Consciousness*, forthcoming.

[5] See Harman, 'Quine on Meaning and Existence, I,' *Review of Metaphysics* **31** (1967—68), 124—151, on the present point pp. 135—137. This splendid work (hereafter just "Harman") is the classic exposition of Quine's views on meaning; my own understanding of analyticity derives very largely from it.

[6] See *W&O*, p. 65, n. 3. I conjecture that all the charges of circularity and emptiness of characterization were really meant to defame the idea that behind philosophers' "intuitions" of analyticity lies a substantive and good theory of meaning that supports the intuitions and has explanatory merit of some sort. I agree with Quine that that idea is false, even though unlike him I believe there is a substantive and good theory of meaning.

[7] *W&O*, Ch. 6; 'Propositional Objects,' in *Ontological Relativity and Other Essays* (New York: Columbia University Press, 1969).

[8] 'Could Propositions Explain Anything?' *Canadian Journal of Philosophy* **3** (1974), 427—434.

[9] *Convention* (Cambridge, MA: Harvard University Press, 1969), pp. 174—177 — a book on lingusitic convention addressed in part to the analyticity question, in which, ironically, Lewis repudiates the understanding of analyticity as truth *by* convention. Though what proposition an English sentence expresses is (obviously) a conventional matter, Lewis insists that the truth at every world of the proposition expressed is not conventional at all but is a mind- and cognition-independent modal fact about the universe (p. 207).

[10] N.B., Lewis does not himself claim that there are any analytic sentences. Also, a response can be hazarded on his behalf. Suppose (waiving cardinality problems) that there is a set of all possible worlds. Any set of worlds is a proposition, and so the set of all worlds is a proposition, the Null or necessary one. But a proposition corresponding to a neatly delineated set of worlds is surely expressible in English, even if there are some sets of worlds that are just too heterogeneous to match single English sentences (which I doubt). Thus, some English sentences express the Null Proposition. Which ones? The standard philosophical examples of trivial verbal truths are the obvious candidates; therefore probably they are analytic. My objections to this argument are (i) that it does nothing to show *what it is* about the trivial verbal truths that makes them express the Null Proposition, and (ii) that in the face of Quinean skepticism we have no reason to think that there are multiple possible worlds distinguished from each other only by "logical" or "conceptual" possibilities that are not nomological possibilities.

[11] Graham Nerlich has recently pointed out that Leibnizian verificationist-*cum*-imaginability arguments against absolute space simply though tacitly assume geometrical theses that hold only for Euclidean space — e.g., that a given figure has similar figures of any size! ('How the Simplest World Can Be Puzzling,' presented to the Center for Philosophy of Science, University of Pittsburgh, April 20, 1989.)

[12] D.M. Armstrong, in *The Nature of Mind and Other Essays* (Ithaca, NY: Cornell University Press, 1981), especially essays 5 and 6.

[13] The original objection was stated most fully on pp. 85—98 of 'Truth by Convention,' in *The Ways of Paradox* (New York: Random House, 1966); the recantation occurred in Quine's Foreword to Lewis' book.

[14] So far as natural languages can themselves be formalized, a Carnapian structure can be imposed; a formal "semantics for" a natural language such as English might contain a box of "meaning postulates" or "nonlogical axioms" intended to exhibit facts of lexical meaning. On the surprisingly complex issue that ensues, see my 'Logical Constants and the Glory of Truth-Conditional Semantics,' *Notre Dame Journal of Formal Logic* **30** (1989), 390—400.

[15] See 'Two Dogmas,' pp. 35—36; 'Carnap and Logical Truth,' in P. A. Schilpp (ed.), *The Philosophy of Rudolf Carnap* (LaSalle, IL: Open Court, 1963), p. 392; and elsewhere.

[16] *Patterns of Discovery* (Cambridge: Cambridge University Press, 1965), pp. 99ff.

[17] See the papers collected in *Mind, Language and Reality* (Cambridge: Cambridge University Press, 1975), especially 'The Analytic and the Synthetic,' 'Is Semantics Possible?,' 'The Meaning of "Meaning"' (pp. 254—257), and 'Dreaming and "Depth Grammar"' (pp. 310—315).

[18] I think, incidentally, that this rejoinder is what Quine has in mind when he makes a somewhat misleading remark on p. 37 of 'Two Dogmas,' calling the analytic/synthetic distinction a "metaphysical article of faith." He means, not that the distinction is a prediction that may or may not come true, but that it is something that is dogmatically believed come what may in the absence of any convincing evidence at all. Pp. 122—127 of 'Carnap and Logical Truth' bear a similar interpretation.

[19] See M. J. Cresswell, 'Adverbs and Events,' *Synthese* **28** (1974), 455—481 (on the present point p. 470), and my 'Logical Constants and the Glory of Truth-Conditional Semantics,' *loc. cit.*

[20] I think he would also want to add exactly what he later says about "explication" generally in Secs. 53 and 54 of *W&O*: that theoretical "definition" is a replacement or substitution, not an analysis or uncovering of a pre-existing conceptual meaning.

[21] See also Harman, pp. 140—141; Quine's remarks ('Carnap and Logical Truth,' p. 113) on the rapidity with which stipulative definitions "fade away"; and again Putnam's many examples (*op. cit.*).

[22] Cf. pp. 71—72 of 'Truth by Convention.'

[23] An earlier stab was made by David Schwayder, in *The Stratification of Behavior* (New York: Humanities Press, 1965).

[24] E.g., Dale Jamieson, 'David Lewis on Convention,' *Canadian Journal of Philosophy* **5** (1975), 73—81; Margaret Gilbert, 'Game Theory and *Convention*,' *Synthese* **46** (1981), 41—94.

[25] It is worth noting in the spirit of Putnam (*op. cit.*) that even if stipulative definition does yield some analytic truths, they certainly are not the ones that are useful to philosophers, being absurdly trivial. No sentence that has ever been put forward as a necessary truth by a philosopher seriously philosophizing is the product of a stipulative definition. Even when philosophers construct elaborate systems of stipulative definitions (Goodman comes to mind here), our interest is not in the definitions themselves, which can have no nontrivial consequences, but rather in the way in which the system connects up to the real world. And this latter way cannot be stipulative.

PART III

FORMAL DEVELOPMENTS

VEIKKO RANTALA

DEFINITIONS AND DEFINABILITY

Traditionally, the word 'definition' means something like explicit definition and, mainly in the philosophy of science, a very limited class of its generalizations. Traditional accounts of definability are often vague and obscure, however, so that it is not always clear what the word stands for. Hence, it is instructive to place definitions in a more formal framework, as we shall see. Such a maneuvre brings forth sophistication which is not practicable in discursive terms — which may sound somewhat paradoxical to philosophers who are accustomed to think of formal methods as yielding intellectual restrictions rather than sophistication. Conversely, there are important pragmatic aspects of definition which cannot be dealt with by using logical tools but rather tools borrowed from the philosophy of language.

In Section 1, I shall outline some well-known views of traditional notions of definition and the distinction often made between nominal and real definition. In Section 2, pragmatic aspects of definitions will be looked at from a point of view of speech act theory. Sections 3—5 are devoted to the modern logical theory of definability, which is usually presented in model-theoretic terms, and in Section 6 some traditional requirements concerning definitions are given a model-theoretic interpretation. Finally, Section 7 studies the methodological and cognitive import of definitions. I shall not attempt any comprehensive survey of traditional or modern theories of definability, but rather focus on some central points which I shall critically study.

1. TRADITIONAL ASPECTS

Whitehead and Russell argue in *Principia Mathematica* that a definition is wholly concerned with symbols and not with what they symbolize and that it is not true or false since it is an expression of a volition and not of a proposition. On the other hand, however, they define a definition as ". . . a declaration that a certain newly-introduced symbol or combination of symbols is to mean the same as a certain other combination of symbols of which the meaning is already known"

J.H. Fetzer, D. Shatz and G. Schlesinger (eds.), *Definitions and Definability: Philosophical Perspectives*, 135—159.

(Whitehead and Russell, 1973, p. 11). Hence they seem to say at the same time that a definition is wholly verbal and that it is not wholly verbal but concerned with meanings as well. They again emphasize the latter aspect as they say that definitions often convey more important information than propositions in which they are used.

It should be clear, in any case, that if a definition is concerned with verbal expressions, it is also concerned with their meanings. A real difficulty that presents itself here is not, however, whether a definition is or is not concerned with meanings but the question of what is intended when it is said that a combination of symbols is to mean the same as a certain other combination of symbols. If meanings are contextual or relative, rather than objective, one has to specify the context or theoretical framework with respect to which meanings are considered. It is clear, of course, that in *Principia* this has been done, but this is not always the case in the traditional discussion of definability.

The notion of definition discussed in *Principia* is sometimes labelled as *nominal definition*, as, for instance, in Cohen and Nagel (1961). According to them, a nominal definition is "... an agreement or resolution concerning the use of verbal symbols. A new symbol called the *definiendum* is to be used for an already known group of words or symbols, the *definiens*. The definiendum is thus to have no meaning other than the definiens" (Cohen and Nagel, 1961 p. 228). A standard example of explicit nominal definitions is the definition of classical implication in terms of negation and disjunction:

(1.1) '*p* implies *q*' is by definition equivalent to 'not-*p* or *q*',

or symbolically, for example:

(1.2) $p \rightarrow q =_{\text{df}} \neg p \vee q.$

Another scientific example mentioned is Comte's invention of the word 'sociology' for the study of human relations in organized group life. Furthermore, it is argued that all explicit definitions in the technique of modern mathematics are nominal (Cohen and Nagel, 1961, p. 232).

According to Cohen and Nagel also, a nominal definition is a resolution and not anything that would be true or false. Hence it cannot be a real premiss of any argument and it does not extend our real knowledge — though it can be useful in scientific inquiry for technical reasons and may clarify our ideas.

In Robinson (1965, p. 16), nominal definition is also characterized as something concerning "words or signs or symbols". The purpose of a nominal definition is to report or to establish the meaning of symbol. This can be done either by saying that its meaning is the same as the meaning of another symbol ("word-word" definition) or by saying that it means a certain thing ("word-thing" definition). Since some nominal definitions are reports, they have a truth value. Hence Robinson's notion of nominal definition extends the notions mentioned in the above.

It is argued in Cohen and Nagel (1961, p. 230) that in a *real definition* the definiens is an analysis of an idea, form, type, or universal symbolized by the definiendum. It defines a word which possesses a meaning independently of the process of definition which equates it with the definiens indicating the structure of the defined entity. It follows that a real definition, unlike a nominal one, is a genuine proposition, true or false. They notice, however, that the distinction between nominal and real definitions is not as sharp as one could think on the basis of their characterization, for even in a verbal definition there usually is some reference to the analysis of the entity in question, and, on the other hand, people may have emotional or other attitudes towards words defined which may confuse the character of definitions.

The distinction which Robinson makes between the two kinds of definition is more advanced. The distinction for him is intentional: 'nominal definition' and 'real definition' refer to *purposes* of definition rather than to methods (Robinson, 1965, p. 16). One's immediate purpose determines to which of the two classes a definition belongs. While the immediate purpose of a nominal definition is to establish the meaning of a symbol, the immediate purpose of a real definition is to do with things ("thing-thing" definition), not symbols *qua* symbols. Nominal definitions are definitions of words whereas real definitions are definitions of things. Robinson (tentatively) suggests, however, that from now on the word 'definition' should, perhaps, be restricted to nominal definitions, i.e., to processes concerning symbols, since the usage of the name 'real definition' has been ambiguous in the literature confusing activities of different kinds under one name (Robinson, 1965, pp. 190—191). He has found twelve such activities, so that, for instance, the analysis of things — which is commonly associated with real definition — is just one of them.

Robinson's view of mathematical definitions is slightly different from the view of Cohen and Nagel (1961). Thus he says, for instance, that definitions in mathematics are often real definitions in the sense of analyses of old ideas — whence they are often statements, true or false and not just declarations concerning meanings of symbols (Robinson, 1965, p. 195).

As we have seen, Whitehead and Russell (1973) and Cohen and Nagel (1961) consider a definition as a syntactic expression in the first place, whereas for Robinson it is first of all an intellectual activity and its other possible senses only are secondary (Robinson, 1965, p. 13). Thus, for example, whether a sentence — or, rather, an utterance — expresses a definition or just a report depends on the corresponding mental state of the speaker. This seems to be another intentional feature in Robinson's notion of definition, and it follows, then, that questions concerning whether something is a definition or which kind of definition are often contextual and matters of interpretation. This suggests that pragmatic dimensions of definability can be studied by considering definitions as speech acts.

2. DEFINITIONS AS SPEECH ACTS

If definitions are speech acts, they can be classified accordingly. Thus, for example, definitions that are real in the sense discussed above would be assertives according to Searle's (1979) classification, and the different activities — or some of them — which Robinson reluctantly mentions as real definitions (see Section 1) might be re-classified by studying their illocutionary force. More generally, if definitions are considered as speech acts, then the whole theoretical machinery of speech acts is available. This would place the traditional discussion in a new perspective. Since, however, the main purpose of the present article is to extend and criticize the traditional discussion by means of logical treatment, I shall restrict myself in this section to some observations concerning pragmatic aspects of definitions which are closely related to speech act theory.

According to Robinson (1965), there are nominal, word-thing definitions with which no truth value can be associated. Such is, for instance, "stipulative definition", the act of assigning an object to a name or a name to an object (Robinson, 1965, p. 59ff.). The corresponding defining utterance, whatever its grammatical form is — there are different forms which are appropriate for the purpose — is not an assertion, but rather a proposal, request, or the like. It obviously is what Austin (1970) calls a *performative* utterance. One of Austin's examples of performative utterances is an act of naming:

(2.1) I name this ship the *Queen Elizabeth.*

By uttering (2.1), one does not describe the christening ceremony, but actually performs the christening, whence, according to Austin, it is not true or false.

Though performative utterances are not true or false, they are, as it is argued by Austin, satisfactory or unsatisfactory, they may succeed or fail. An utterance may suffer from various "infelicities" which arise if certain rules are broken. The rules constraining utterances are pragmatic rather than logical, and they are due to the conventional nature of language and the contextual and intentional character of utterances. First, the alleged social convention on which one relies when performing an act must exist and must be accepted by relevant institutions or communities. The performance of (2.1) fails to name the ship if it is part of a procedure which deviates from the accepted ceremony of christening ships. Second, the context in which an utterance is performed must be acceptable. (2.1) fails, for instance, if a person not authorized for the purpose performs the act according to the formally correct procedure.

Third, since performative utterances are often intentional — stipulative definitions, in particular, are intentional — it is questionable whether a symbol will really become defined or a thing named if an utterance is not sincere or if there is a contradiction between the accepted, conventional interpretation of the utterance and the speaker's intention. If the utterance is not sincere and hence does not express the intentional state the speaker actually has, then it is obvious that there is a kind of infelicity in Austin's sense. On the other hand, however, it seems to depend on the strength of social constraints whether or not the definition is successful in such a case. If, for instance, I insincerely utter (2.1) in proper circumstances and in accordance with accepted rules, and as the authorized person, then it is obvious that I have named the ship the *Queen Elizabeth* since in this case the social constraints in question are powerful enough to overrule my insincerity.

All these cases of infelicity are somewhat inconvenient for Robinson's view that definitions are primarily mental activities. If an utterance is syntactically appropriate, then it seems to follow from what we have said above that in proper circumstances it is its syntactic form rather than the utterer's intention that makes it a definition, that is, it is the syntactic form on the basis of which it is generally accepted as a definition. Similar reservations concern Robinson's view that the distinction between nominal and real definition is intentional. In scientific contexts, the intention of defining something is often indicated by syntactical devices, as in (1.1) and (1.2), above, but in formal languages it is the syntactic form which is crucial.

3. THE LOGICAL ACCOUNT

The syntactic form of explicit definition, the most important kind of definition, is identity or equivalence in some sense. Since in Robinson (1965) the notion of definition is very comprehensive, even explicit definitions have no fixed syntactic form and the question of their form is in fact inessential. Whether or not a verbal expression represents a definition or a definition of a given kind depends on the respective rules. This pragmatic conception is appropriate so far as definitions in natural language are concerned, but it is not appropriate for formal languages. Pragmatic dimensions are suppressed when formal ones are emphasized, and syntactic form becomes essential. It seems, however, that some pragmatic aspects can be illuminated by using model-theoretic distinctions. More importantly, distinctions can be made which are not clear or possible without concepts of logical semantics. Such is, for example, the distinction between global and local definability. It is not always clear whether in traditional accounts a given definition is assumed to be contingent, to apply only to 'actual' objects, or whether they are thought of as being of a more necessary nature, applicable in 'all possible worlds' — even though some attention has been paid to that distinction when real and nominal definitions have been discussed. Furthermore, logical semantics makes it possible to study the semantic import of different kinds of definability.

I shall first make an excursion to some standard notions and notation of model theory, assuming, however, that the reader is more or less familiar with them. I shall mainly restrict the discussion to the familiar first-order predicate logic, that is, elementary logic, and only occasionally refer to other logics.

A similarity type or vocabulary τ is a set of non-logical constants, that is, predicate and function symbols and individual constants. If L is the elementary, first-order language of type τ and P an arbitrary non-logical constant not in τ, then the expanded language of type $\tau \cup \{P\}$ will be denoted by '$L(P)$'. This notation is generalized in an obvious way for any number of any new constants. For simplicity, I shall mainly assume that P is a unary predicate symbol. The assumption could be dispensed with; it would mainly cause some notational inconvenience (and some additional conditions concerning definitions of function symbols and individual constants; see Suppes, 1957). I shall consider the question of what it means to say that P is definable in terms of τ. Hence we are interested in relations of formulas of languages L and $L(P)$. Logical constants, connectives, and quantifiers will be denoted as usual, deducibility by '$\vdash$', and formulas of any language by lower-case Greek letters. If the free variables of a formula ϕ are among $x_1, \ldots, x_n$, we may indicate it by writing '$\phi(x_1, \ldots, x_n)$'. A sentence is a formula with no free variables. A (formal) theory in a language is a set of sentences of the language closed under deduction.

That a sentence ϕ, or a set of sentences Σ, is true in a model $\mathbf{M}$, that is, that $\mathbf{M}$ is a model of ϕ or Σ, will be denoted, as usual, by '$\mathbf{M} \models \phi$', or '$\mathbf{M} \models \Sigma$', respectively. If P is a unary predicate symbol and $\mathbf{M}$ is a model for the language L (model of type τ), i.e., a model in which the non-logical constants of the language are interpreted, an arbitrary expansion of $\mathbf{M}$ to the language $L(P)$ will be denoted by '$(\mathbf{M}, X)$', where X is a subset of M, the domain of $\mathbf{M}$. Thus P, that was added to τ, is interpreted as X. This notation is again generalized in an obvious way.

For aesthetic reasons, the following abbreviations will be used: $\mathbf{a}$, $\mathbf{c}$, $\mathbf{x}$ for '$a_1, \ldots, a_m$', '$c_1, \ldots, c_m$', '$x_1, \ldots, x_m$', respectively; $\mathbf{z}$ for '$z_1, \ldots, z_k$'; and $\exists \mathbf{x}$, $\forall \mathbf{z}$ for $\exists x_1 \ldots \exists x_m$, $\forall z_1 \ldots \forall z_k$, respectively.

With these notions at hand, various formal definability concepts can be introduced. I shall consider explicit definability and some weaker notions and only summarize, with no proofs, some known results of elementary logic which relate syntactic and semantic aspects of these notions to each other. (For a more detailed discussion and proofs, see the works referred to below and Rantala, 1977a.) A syntactic definition means here a sentence of a certain form in the formal object language, that is, in this section we shall not consider syntactic definitions in the metalanguage. Furthermore, since it does not make much sense to consider definitions *per se*, we shall consider such notions as definability in a theory and definability in a model.

Let us assume that τ is an arbitrary vocabulary, L the elementary (i.e., first-order) language of type τ, P a unary predicate symbol not in τ, and $T(P)$ an elementary theory (i.e., formulated in elementary logic) in the language $L(P)$. As remarked above, the indicated restriction concerning P is only stated for expositional simplicity. We consider first the notions of explicit and implicit definability and their semantic counterpart. One of the most important model-theoretic results of elementary logic is the following, Beth's Theorem (which generalizes a result of Tarski, 1956):

THEOREM 3.1 (Beth, 1953). The following conditions are equivalent.

(i) There is a formula $\phi(x)$ of L such that

$$T(P) \vdash \forall x(P(x) \leftrightarrow \phi(x)).$$

(ii) If Q is a unary predicate symbol not in τ and $T(Q)$ the theory in $L(Q)$ obtained from $T(P)$ by substituting Q for P in $T(P)$, then

$$T(P) \cup T(Q) \vdash \forall x(P(x) \leftrightarrow Q(x)).$$

(iii) For every model $\mathbf{M}$ for L, $\mathbf{M}$ has at most one expansion $(\mathbf{M}, X)$ to the language $L(P)$ such that $(\mathbf{M}, X)$ is a model of $T(P)$.

Here X is of course a subset of the domain M of $\mathbf{M}$. The condition (i) says that P is *explicitly definable* in $T(P)$ (in terms of τ), i.e., that an explicit definition of P is derivable from the theory. This notion covers, of course, both of the two methodological situations: that the theory has been developed by adding an explicit definition to previous axioms (and thus expanding the language) and that the theory is given and an explicity definition is derived only afterwards. (ii) expresses the fact that P is *implicitly definable* in $T(P)$. (iii) is the semantic counterpart of (ii), that is, their equivalence is immediate and independent of the assumption that we consider elementary logic here. If (iii) holds, P will be here called *identifiable* in $T(P)$. (Notice that the term 'identifiable' has different meanings in the literature; see Rantala, 1977b.) The equivalence of explicit and implicit definability in elementary logic, or, rather, that implicit definability implies explicit definability, is a deep result which is a consequence of Craig's Interpolation Theorem (see, e.g., Chang and Keisler, 1973) and it does not hold for all logics. It is one of the characteristic properties of elementary logic, and it has been pointed out that well-known extensions and deviations of elementary logic do not generally have this property (see, e.g., Makowski, Shelah, and Stavi, 1976). The same obviously holds for the results below.

The model-theoretic condition (iii) of Theorem 3.1 is in fact quantitative in the sense that it states a restriction on the number (cardinality) of acceptable interpretations of P once the interpretations of the symbols in τ are fixed: there cannot be more than one such interpretation. The next two definability notions also impose such quantitative restrictions, but in a generalized sense. That the term 'definability' can be justifiably applied to these notions is not obvious as we shall see, but one justification will be provided later in Section 6.

THEOREM 3.2 (Kueker, 1970). The following conditions are equivalent.

(i) There are formulas $\sigma(\mathbf{x})$ and $\phi_i(\mathbf{x}, y)$ $(i = 1, \ldots, n)$ of L such that

(a) $$T(P) \vdash \exists \mathbf{x} \sigma(x);$$

(b) $$T(P) \vdash \forall \mathbf{x}(\sigma(\mathbf{x}) \rightarrow \bigvee_{1 \leqslant i \leqslant n} \forall y(P(y) \leftrightarrow \phi_i(\mathbf{x}, y))).$$

(ii) For every model $\mathbf{M}$ for L, $\mathbf{M}$ has at most n expansions $(\mathbf{M}, X)$ to the language $L(P)$ such that $(\mathbf{M}, X)$ is a model of $T(P)$.

In Hintikka (1972), P is said to be *n-foldly identifiable* in $T(P)$ (in terms of τ) if (ii) holds and *finitely identifiable* in $T(P)$ if it is n-foldly identifiable for some (finite) n. We shall here call P *finitely definable* in $T(P)$ if (i) holds for some n. Thus, in elementary logic finite identifiability and finite definability are equivalent notions. If, in particular, $n = 1$, then P is explicitly (and implicitly) definable. The explicit definition which is then obtained from the conditions (a) and (b) is as follows:

$$\forall y(P(y) \leftrightarrow \exists \mathbf{x}(\sigma(\mathbf{x}) \wedge \phi_1(\mathbf{x}, y))).$$

In the next theorem, the number of acceptable expansions can be infinite. Let '$|M|$' denote the cardinality of M:

THEOREM 3.3 (Chang, 1964; Makkai, 1964). The following conditions are equivalent.

(i) There are formulas $\phi_i(\mathbf{x}, y)$ $(i = 1, \ldots, n)$ of L such that

$$T(P) \vdash \bigvee_{1 \leqslant i \leqslant n} \exists \mathbf{x} \forall y(P(y) \leftrightarrow \phi_i(\mathbf{x}, y)).$$

(ii) For every infinite model $\mathbf{M}$ for L, $\mathbf{M}$ has at most $|M|$ expansions $(\mathbf{M}, X)$ to the language $L(P)$ such that $(\mathbf{M}, X)$ is a model of $T(P)$.

In Hintikka (1972), P is called *restrictedly identifiable* in $T(P)$ (in terms of τ) if (ii) holds. In that case, a fixed interpretation of the symbols in τ in an infinite domain M leaves at most $|M|$ choices open for interpreting P in order to obtain a model of $T(P)$ — even though the number of all possible interpretations is $2^{|M|}$. P is *restrictedly definable* in $T(P)$ if (i) holds.

The next theorem states a criterium for a term being definable in the models of a theory. It is said that P is *definable in a model* for $L(P)$ (in terms of τ) if an explicit definition $\forall x(P(x) \leftrightarrow \phi(x))$, where $\phi(x)$ is a formula of L, is true in the model.

THEOREM 3.4 (Hintikka and Tuomela, 1970). The following conditions are equivalent.

(i) There are formulas $\phi_i(x)$ $(i = 1, \ldots, n)$ of L such that

$$T(P) \vdash \bigvee_{1 \leqslant i \leqslant n} \forall x(P(x) \leftrightarrow \phi_i(x)).$$

(ii) P is definable in every model of $T(P)$.

(iii) P is explicitly definable in every complete consistent extension of $T(P)$.

It is said that P is *piecewise definable* in $T(P)$ (in terms of τ) if (i) holds. If P is piecewise definable, then P is definable in every model of the theory, but the definiens may be different in different models. This notion is again a generalization of the notion of explicit definability, which results if $n = 1$. On the other hand, it is implied by finite definability in the special case that the 'parameters' $\mathbf{x}$ can be dispensed with (see Kueker, 1970). The conditions (i)—(ii) also have an important algebraic equivalent, as shown by Svenonius (1959).

In actual theory constructions, *conditional definitions* are often used rather than explicit ones, that is, defined notions are not assumed to be uniquely applicable to all elements of the domains in questions whence some preconditions are needed. Consider, for example, the ordinary definition of division in arithmetic where a condition must be stated which excludes division by zero. Other examples are not hard to come by — although required conditions are not always explicitly stated.

Conditional definitions also play an important historical role in the philosophy of science, in logical empiricists' discussion concerning the relation of 'theoretical terms' and 'observational terms' in scientifc theories. For example, it was observed by Carnap (1936—37) that certain theoretical terms, such as dispositional predicates, are not explicitly definable — explicit definablity was initially required by logical empiricists — and generally not even conditionally definable in terms of observational terms in an intuitively adequate way if theories are assumed to be reconstructed in elementary logic. This drawback can be taken care of if elementary logic is extended by adding subjunctive conditionals — which results, however, in difficult problems concerning formal semantics. (See e.g., Przelecki, 1969; Tuomela, 1973; Suppe, 1977, for the problem concerning the definability of theoretical terms; and Goodman, 1947, and Lewis, 1973, for subjunctive conditionals.) This characterization problem is just one indication of the fact that the expressive power of elementary logic is too weak for many metascientific purposes. Other examples, which are rather different, however, will be offered below.

In our present, slightly simplified logical framework, it is said that P is *conditionally definable* in $T(P)$ (in terms of τ) if a conditional definition is derivable from $T(P)$:

$$T(P) \vdash \forall x(\sigma(x) \rightarrow (P(x) \leftrightarrow \phi(x))).$$

Here σ and ϕ are formulas of L. The semantic import of conditional definability can be readily understood. For the cases of conditional definability where our simplification will not do — most mathematical definitions, for instance, are such — that is, where instead of a unary predicate symbol we have to have nonlogical constants of other kinds, some conditions for proper definitions must be added (see Suppes, 1957).

It is not possible to make here any exhaustive overview of the different notions of definition which are mathematically, logically, or methodologically important and representable in formal logic. For other notions, I can only refer to standard textbooks.

4. UNDEFINABILITY AND UNCERTAINTY

As we have seen, if P is explicitly definable in $T(P)$, then each interpretation of τ uniquely determines the interpretation of P in the sense that every model for L has at most one expansion to a model of $T(P)$. Conversely, if P is not explicitly definable in $T(P)$, the theory leaves some freedom to interpret P. This freedom can also be construed as a kind of uncertainty concerning the interpretation of P. It means a kind of uncertainty for a person who is studying the theory and its models if he only knows how the constants in τ have been interpreted. Only if P is explicitly defined in the theory, may the person know how P has been interpreted in an arbitrary model of the theory *on the basis* of knowing how these other constants have been interpreted. The other kinds of definability do not completely remove this uncertainty, but some of them reduce it as we can see from Theorems 3.2—3.4.

It depends on the methodological standpoint whether undefinability is called freedom or uncertainty. The latter name is motivated, for example, if one is studying a theory and its models by stepwise processes during which one is gathering more and more information concerning how P and τ are syntactically or semantically related. It has been pointed out in Hintikka and Tuomela (1970) and Hintikka (1972) that by using an appropriate method of elementary logic — the method of distributive normal forms — a stepwise syntactic process can be defined by means of which *global uncertainty* can be studied, that is, uncertainty concerning the interpretation of P in all models of the theory. One's initial uncertainty concerning P may be reduced during the process, and at some point completely removed if P is explicitly definable in $T(P)$. If, on the other hand, P is definable in one of the senses indicated in Theorems 3.2—3.4, it will also become apparent at some point of the process. Conversely, if certain distinctive syntactic features are displayed in the course of the process, one can infer that P is definable in $T(P)$ in one of these senses. Thus, *global* definability and the reduction of global uncertainty go hand in hand.

The method in question is based on Hintikka's notion of constituent, a special kind of formula, which can be used to represent various properties of elementary theories. In Hintikka and Rantala (1975) and Rantala (1977a), this syntactic method is combined with model-theoretic ones to study *local* uncertainty, uncertainty concerning interpretation in single models. By these methods local definability results are obtained, that is, results concerning (finitary and infinitary) definability in a model. The notion of definability in a model is generalized for the other forms of definition with which we have become acquainted in the above, and in some of the mentioned results formulas occurring in definitions are infinitary, containing infinite conjunctions or disjunctions, and hence do not belong to elementary logic.

5. MODEL-THEORETIC DEFINABILITY

In model theory, questions concerning the expressive power of various logics extending elementary logic have been extensively studied. It is often important for metamathematical purposes to know whether given classes of mathematical structures, that is, of models, can be characterized in some logic in some well-defined sense. If **L** is a logic and $\models$ is now the truth relation in **L**, and K is a class of **L**-models, then to say that K is characterizable in **L** means in the most straightforward sense that there is a sentence ϕ (or possibly an appropriate set of sentences) of **L** such that K is the class of all models in which ϕ is true, i.e., such that the following holds for all models **M**:

$$\mathbf{M} \in K \text{ if and only if } \mathbf{M} \models \phi.$$

This equivalence expresses an explicit definition in the metalanguage of **L** (derivable in the metatheory of **L**), and hence it is natural to say that K is *definable* in **L**. The expressive power of elementary logic is weak for metamathematical purposes in the sense that not all important classes of mathematical structures, hence not all important mathematical notions, are definable in elementary logic. Such are, for example, the class of all finite sets (that is, the notion of finiteness), the class of all structures isomorphic to the standard model of arithmetics of natural numbers, and the class of all well-ordered sets (that is, the notion of well-ordering), to mention some of the most important cases.

Questions concerning the characterization of classes of models in different logics belong to what in mathematical logic is called *definability theory* — which covers large and divergent areas of mathematical logic (see Barwise, 1975). Our earlier discussion concerning the definability of P in $T(P)$ is of course part of that general definability theory, and such results as Theorems 3.1—3.4 have often been construed as indicating that there obtains a balance between expressive power and mathematical elegance of elementary logic. Similar results need not hold for extensions or deviations of elementary logic, as we already remarked in the above.

We may see now from this and preceding sections that as soon as definitions are studied in the model-theoretic framework, distinctions are made which are completely beside the question when they are studied in natural or informal scientific language. On the other hand, it is evident why such pragmatic considerations as the ones discussed in Section 2, above, are not very relevant in the logical theory of definability. For instance, whether a formal expression indicates a definition is not so much dependent on the intentional state of the 'speaker'; it is determined by logical conventions and rules.

6. CONDITIONS ON DEFINITIONS

Definitions are traditionally assumed to satisfy various requirements in order to be acceptable. Some of them are methodological, such as the rules that a definition must state the essence of the defined object and be given *per genus et differentiam* and that a definition must not be expressed in an obscure way, or logical, such as the rules that a definition must not be expressed negatively if it can be expressed positively and that the definiendum must be equivalent to the definiens. Aristotle's work is usually considered as the historical origin of these rules. There also exist other conditions — some of which are of pragmatic character — in the literature which are of later origin. (For a comprehensive discussion of various historical conditions, see Robinson, 1965.)

In this section, the requirements of *eliminability* and *noncreativity* will be studied and generalized. These requirements can be formulated in the context of formal and mathematical languages, rather than for natural language. It is argued in *Principia Mathematica* that a definition is theoretically unnecessary in the sense that one might always use the definiens and dispense with the definiendum (Whitehead and Russell, 1973, p. 11). This condition seems to amount to eliminability which in our formal framework is formulated as follows. P is *eliminable* in $T(P)$ (in terms of τ) if for each formula ϕ of $L(P)$, there is a formula ϕ' of L (i.e., formula not containing P) such that

$$(6.1) \qquad T(P) \vdash \phi \leftrightarrow \phi'.$$

This notion is typically applied to cases in which a new term is introduced by means of a definition when a theory is being constructed, that is, in which $T(P)$ is the set of all sentences of $L(P)$ deducible from a set $\Sigma \cup \{\delta\}$, where Σ is a set of sentences of L (e.g., axioms and definitions of $T(P)$ or another theory) and δ is a definition (of some kind) of P in terms of τ.

(6.2) $T(P) = Cn(\Sigma \cup \{\delta\})$.

Then it is often required that the definition δ must be such that P is eliminable in $T(P)$. The notion of eliminability is here defined more generally, however, so that it applies to arbitrary methodological situations.

If P is eliminable in $T(P)$, then for each formula of $L(P)$ there is a formula of L such that the two formulas are equivalent in $T(P)$ — but they are not in general logically equivalent. In this sense the eliminability of P is only contextual. As can be expected, eliminability and explicit definability are equivalent notions. If P is explicitly definable in $T(P)$, then ϕ' is obtained by replacing each subformula $P(x)$ of ϕ (where x is an arbitrary variable of the language) by the respective definiens. Conversely, if P is eliminable, then an explicit definition of P is obtained by choosing $P(x)$ for ϕ in (6.1). It follows, then, by Beth's Theorem, that the eliminability of P is equivalent to the condition that each model for L has at most one expansion to $L(P)$ which is a model of $T(P)$ (cf. Theorem 3.1):

THEOREM 6.1. The following conditions are equivalent:

(i) P is eliminable in $T(P)$;
(ii) P is explicitly definable in $T(P)$;
(iii) For every model $\mathbf{M}$ for L, $\mathbf{M}$ has at most one expansion $(\mathbf{M}, X)$ to the language $L(P)$ such that $(\mathbf{M}, X)$ is a model of $T(P)$.

That explicit definability implies eliminability is in fact a precise way to say that explicit definability has the following important methodological consequence, which is much discussed in the literature, but not always precisely expressed: $T(P)$ is *reducible* to a theory in L (whose vocabulary is simpler) — in the constructive case (6.2) to the theory $Cn(\Sigma)$. *Principia Mathematica* is a well-known attempt to reduce mathematical theories to logical ones in this sense, and there are many other similar reduction results in mathematics, mathematical logic, set theory, and methodology of science. The notion of reduction relation which is connected with explicit definability is in fact a special case of a more general notion. (For the general notion, see, e.g., Pearce and Rantala, 1983, 1984; Pearce, 1987.)

Is a similar concept of eliminability possible for definitions in natural language, that is, if we consider common speech instead of scientific and logical languages? Since it is difficult to specify different levels of language and to define the notion of linguistic context in natural language, the best thing one can do here is to define the eliminability of a symbol S as the feature that for each expression A in which S occurs, there is another one, B, in which it does not occur and which has the same meaning as A or which is synonymous to A. The notions of meaning and synonymy are, however, notoriously vague and confused notions and they are criticized by several philosophers, Quine (1953) in the first place. Furthermore, the question how B is obtained from A is perplexing. If it is assumed, for example, that S is (explicitly) definable by means of a word-word definition (cf. Section 1) and that B is obtained by replacing S by the definiens, we get into well-known troubles with intensional contexts, e.g., if A is a propositional attitude phrase (see, e.g., Quine, 1953; Cresswell, 1973; Saarinen, 1979), and, more generally, with linguistic contexts which are not appropriately chosen. That is, only if relativized to appropriate, sufficiently narrow fragments of natural language does the notion of eliminability make sense, but, on the other hand, the task of delineating relevant fragments of common speech is as problematic as the task of defining the notion of synonymy or the sameness of meaning. The notion is fully relevant only for languages which are more technical and extensional than natural language. The same holds for the notion of noncreativity, which will be studied next.

Noncreativity is defined for theories of the form (6.2) where δ is now any sentence of $L(P)$ which is used to extend Σ, i.e., to construct $T(P)$, but which we may here think of as a definition. It is said that the sentence δ is *noncreative* in $T(P)$ if any theorem of $T(P)$ not containing P can already by deduced from Σ alone, that is, if for any sentence ϕ of L such that $T(P) \vdash \phi$, it also holds that $\Sigma \vdash \phi$. In other words, δ is noncreative if $T(P)$ is a *conservative extension* of Σ.

If δ is noncreative, no new consequences in the language L can be obtained by means of δ, but it does not mean that no new consequences at all could be created by adding δ to Σ. New consequences are, of course, obtained, but they contain P, and they need not be reducible to L if P is not also eliminable, that is, explicitly definable.

Noncreativity can also be expressed in semantic terms. Let us say that P is *Ramsey-eliminable* in $T(P)$, if every model of Σ (which is a model for L) is expandable to a model of $T(P)$. There exist different formulations of Ramsey-eliminability, which has in fact played an important methodological role in the recent philosophy of science (see Tuomela, 1973). The following theorem — whose converse does not hold in general — is a straightforward consequence of the completeness of elementary logic:

THEOREM 6.2. If P is Ramsey-eliminable in $T(P)$, then δ is noncreative in $T(P)$.

The following theorem shows that noncreativity is in a sense a weaker notion than eliminability. The theorem will be proved here since the result is not, perhaps, well-known. The theorem justifies — in a weak sense — our terminology in Section 3, above, where we generalized the notion of definability.

THEOREM 6.3. The global definitions mentioned in Section 3 are noncreative.

Proof. It is sufficient to consider restricted definability. Thus we assume that $T(P)$ is as in (6.2), where δ is now of the form

$$\bigvee_{1 \leqslant i \leqslant n} \exists \mathbf{x} \forall y (P(y) \leftrightarrow \phi_i(\mathbf{x}, y))).$$

We show that in this case P is Ramsey-eliminable in $T(P)$, whence noncreativity followings. Let $\mathbf{M}$ be a model of Σ and let $\phi_i(\mathbf{x}, y)$ be any of the formulas indicated above. Let $\mathbf{a}$ be arbitrary elements of the domain M of $\mathbf{M}$ and let $\mathbf{c}$ be new individual constants (not in L) which will be interpreted as $\mathbf{a}$, respectively. Furthermore, let X be the set of all elements of M which satisfy the formula $\phi_i(\mathbf{c}, y)$ in $(\mathbf{M}, \mathbf{a})$ (X may be empty). It follows that

$$(\mathbf{M}, X, \mathbf{a}) \models \forall y (P(y) \leftrightarrow \phi_i(\mathbf{c}, y)),$$

and hence

$$(\mathbf{M}, X) \models \exists \mathbf{x} \forall y (P(y) \leftrightarrow \phi_i(\mathbf{x}, y)),$$

that is,

$$(\mathbf{M}, X) \models \delta.$$

From this it follows that

$$(\mathbf{M}, X) \models \Sigma \cup \{\delta\},$$

whence $(\mathbf{M}, X)$ is also a model of $T(P)$, and so P is Ramsey-eliminable.

Hence all the definitions considered in Section 3 can be thought of as proper definitions in the weak sense that they do not admit new consequences in L. With the exception of explicit definition, they do not meet the criterium of eliminability, as we saw, but their image as proper definitions can be strengthened by showing that they satisfy certain weaker eliminability criteria: they admit eliminability in some local or contextual senses (cf. Rantala, 1977a). We define first a notion of local eliminability in an obvious way. P is *eliminable in a model* $\mathbf{M}$ for $L(P)$ if for every formula $\phi(\mathbf{z})$ of $L(P)$, there is a formula $\phi'(\mathbf{z})$ of L such that

$$\mathbf{M} \models \forall \mathbf{z}(\phi(\mathbf{z}) \leftrightarrow \phi'(\mathbf{z})).$$

To each of the above forms of definability there corresponds a weak notion of eliminability in an obvious way. Let us start with restricted definability. We say that P is *restrictedly eliminable* in $T(P)$ if for every formula $\phi(\mathbf{z})$ of $L(P)$, there are formulas $\phi_i(\mathbf{x}, \mathbf{z})$ $(i = 1, \ldots, n)$ such that

$$T(P) \vdash \bigvee_{1 \leqslant i \leqslant n} \exists \mathbf{x} \forall \mathbf{z}(\phi(\mathbf{z}) \leftrightarrow \phi_i(\mathbf{x}, \mathbf{z})).$$

That this notion corresponds to restricted definability is shown by the following theorem:

THEOREM 6.4. The following conditions are equivalent.

(i) P is restrictedly definable in $T(P)$.
(ii) P is restrictedly eliminable in $T(P)$.

Proof. Assume that (i) holds. Then there are formulas $\psi_i(x, y)$ $(i = 1, \ldots, n)$ of L such that

$$T(P) \vdash \bigvee_{1 \leqslant i \leqslant n} \exists \mathbf{x} \forall y(P(y) \leftrightarrow \psi_i(\mathbf{x}, y)).$$

If $\mathbf{M}$ is a model of $T(P)$, then for some $i = 1, \ldots, n$, it holds that

$$\mathbf{M} \models \exists \mathbf{x} \forall y(P(y) \leftrightarrow \psi_i(\mathbf{x}, y)),$$

whence for some elements $\mathbf{a}$ of M, the domain of $\mathbf{M}$,

$$(\mathbf{M}, \mathbf{a}) \models \forall y(P(y) \leftrightarrow \psi_i(\mathbf{c}, y)),$$

where **c** are new individual constants, interpreted as **a**, respectively. Thus, P is definable in the model $(\mathbf{M}, \mathbf{a})$ for the language $L(\mathbf{c})$ (in terms of $\tau \cup \{\mathbf{c}\}$). It follows that if $\psi(\mathbf{z})$ is an arbitrary formula of $L(P)$, there is a formula $\psi_i(\mathbf{c}, \mathbf{z})$ of $L(\mathbf{c})$ such that

$$(\mathbf{M}, \mathbf{a}) \models \forall \mathbf{z}(\phi(\mathbf{z}) \leftrightarrow \phi_i(\mathbf{c}, \mathbf{z})),$$

$$\mathbf{M} \models \exists \mathbf{x} \forall \mathbf{z}(\phi(\mathbf{z}) \leftrightarrow \phi_i(\mathbf{x}, \mathbf{z})).$$

Since for each model $\mathbf{M}$ of $T(P)$, this holds of some $\psi_i(\mathbf{x}, y)$ $(i = 1, \ldots, n)$, it follows that there are formulas $\psi_i(\mathbf{x}, \mathbf{z})$ $(i = 1, \ldots, n)$ such that

$$T(P) \vdash \bigvee_{1 \leqslant i \leqslant n} \exists \mathbf{x} \forall \mathbf{z}(\phi(\mathbf{z}) \leftrightarrow \phi_i(\mathbf{x}, \mathbf{z})),$$

whence (ii) holds.

Conversely, it is obvious that (ii) implies (i).

Restricted eliminability is a kind of contextual eliminability. If P is restrictedly eliminable in $T(P)$, then — as can be readily seen from the above proof — for each formula of $L(P)$ and for each model $\mathbf{M}$ of $T(P)$, P can be eliminated in a relevant expansion $(\mathbf{M}, \mathbf{a})$ of $\mathbf{M}$, in terms of τ plus some new *parameters* **c**. But with piecewise definability a notion of contextual eliminability without parameters can be associated, as can be expected. P is called *piecewise eliminable* in $T(P)$ (in terms of τ) if for every formula $\phi(\mathbf{z})$ of $L(P)$, there are formulas $\phi_i(\mathbf{z})$ $(i = 1, \ldots, n)$ of L such that

$$T(P) \vdash \bigvee_{1 \leqslant i \leqslant n} \forall \mathbf{z}(\phi(\mathbf{z}) \leftrightarrow \phi_i(\mathbf{z})).$$

The proof of the following theorem is similar to the proof of the previous one:

THEOREM 6.5. The following conditions are equivalent.
(i) P is piecewise definable in $T(P)$.
(ii) P is piecewise eliminable in $T(P)$.
(iii) P is eliminable in every model of $T(P)$.
(iv) P is eliminable in every complete consistent extension of $T(P)$.

It should be obvious now, how the eliminability notions corresponding to finite and conditional definability are defined. Their exact definitions are left to the reader.

7. WHY ARE DEFINITIONS NEEDED?

If a definition is noncreative and the term it introduces eliminable, is the definition worth the effort? The question is more relevant about definitions in formal and scientific languages than in natural language since, as we saw, eliminability and noncreativity are notions that are not straightforwardly applicable to definitions in common speech. Furthermore, it is more relevant about explicit definition than the other kinds of definition discussed above since eliminability in the strict sense is not true of the latter. Since eliminability in appropriate local and contextual senses holds for the other kinds, the question is of some relevance in their cases as well. On the other hand, as it has been pointed out in the extensive literature of the philosophy of science, introducing new 'theoretical' terms to a scientific theory is *in general* of great methodological, pragmatic, and heuristic importance (see, e.g. Przelecki, 1969; Tuomela, 1973).

Many authors have posed this question and given rather similar answers. Whitehead and Russell (1973, pp. 11—12) state the core of the answers as they say, firstly, that though definitions are "theoretically superfluous", they nevertheless "... often convey more important information than is contained in the propositions in which they are used." This follows from the fact that when a definition is chosen, it shows us that its definiens is considered important, and from the fact that the definiendum is often more or less familiar to us, so that the definition, being a result of an analysis of a common idea, "... may therefore express a notable advance." A definition may, for instance, make a vague idea precise, as was made by Cantor's definition of continuum. Secondly, there is a practical or economical dimension: "... if we introduced no definitions, our formulas would very soon become so lengthy as to be unmanageable ...."

What is said in *Principia* about the nature of information definitions may convey is of course very true, but something must be added to it. According to Frege (1884), conclusions we draw from a fruitful definition extend our knowledge. This remark seems to refer to a kind of information which is different from the kind of information mentioned in *Principia*, and it is exactly opposite to the view held by Cohen and Nagel (1961) — as far as nominal definitions are concerned.

What does it mean to say that a definition is fruitful and that conclusions drawn from it extend our knowledge? According to Frege, a definition is fruitful only if it draws boundary lines which were not previously given at all. If, for example, the term defined is a unary predicate expression, then the definiens describes a collection of individuals in each world in which the definition is assumed to be true. In this sense, the definition draws "boundary lines" in each such world. Therefore, one way to construe Frege's notion of fruitfulness is to say that a definition is fruitful if it makes a new classification of objects — new, e.g., in the sense that it was not known before or in the sense that it was not made, or that it was not known that it was made, by the earlier theory to which the definition was added. This kind of knowledge seems to be close to the kind of knowledge Robinson (1965, p. 33) means as he says that a word-thing definition ". . . may give us new knowledge of the thing in that it abstracts it and sets it off from the rest of the world in a way which we probably had not done before . . . ." In this way a definition may ". . . teach the contents of the world and the variety of human ideas." As an example Robinson provides the definition of 'derivative' in mathematics, whereby a student learns this mathematical notion.

This kind of fruitfulness can be model-theoretically characterized by applying to definability Hintikka's notions of distributive normal form and semantic information. The characterization seems to lead to the conclusion made in Hintikka and Tuomela (1970) that definitions can be fruitful in the Fregean sense even though they are noncreative and the terms they introduce eliminable (see also Rantala, 1977a). It appears, furthermore, that the activities which are needed — in the framework of distributive normal forms — to come upon new boundaries and to make them significant to us are analogous to the activities needed in logical deductions, as they are characterized in Hintikka (1973) (for the latter activities, see also Rantala and Tselishchev, 1987). Hence, it seems to follow from Hintikka's theory of semantic information that there is an analogy between the ways in which information is obtained from fruitful definitions and from fruitful logical argumentation. This analogy can be considered as one aspect of the parallelism, emphasized by Tarski (1956), between the theory of definability and the theory of deduction.

Let us observe in passing that a definition is often used to introduce new objects — not only new boundary lines — as, for example, when a domain of known individuals is extended in science or mathematics (see, e.g., Przelecki, 1977). Well-known mathematical examples are provided by various extensions of number domains. However, in what sense one considers the defined objects 'new', e.g., whether one thinks of the definition as introducing only new boundary lines or even ontologically new objects, depends on one's philosophical attitude. Philosophically this is a choice between realism and constructivism, but the choice may have implications concerning what methodological and formal procedures one chooses. It is clear that the kinds of information obtained in the two cases are different.

As observed in *Principia*, the main economical gain obtained by introducing definitions to a theory is that they reduce complexity. This is a more obvious advantage of definitions. It is clear, for example, that if a complex expression of natural language or scientific language is replaced by a simple term, it may be of a considerable cognitive and practical value. This is what Hempel (1952, p. 17) means as he says that a formulation of theories solely in terms of primitives may become so involved as to be unintelligible. One aspect of the reduction of complexity is the following, as Hintikka and Tuomela (1970) point out by using Hintikka's notion of semantic information. Consider a first-order theory $T(P)$ as in (6.2), above. Even if the definition δ is noncreative, so that the theorems of $T(P)$ which belong to L can be deduced from Σ already, it is possible that within $T(P)$ they can be deduced in a simpler way. The complexity of deductions may thus be reduced. Such an advantage can be expected if the definition ". . . exhibits a genuine quantificational analysis of the definiendum, in the sense that its definiens has one or more (irreducible) layers of quantifiers . . ." (Hintikka and Tuomela, p. 317). Though this observation concerns the formal theory of definition, it suggests that a similar gain is obtainable from informal definitions also. It is not difficult to find examples — e.g., by studying definitions in mathematics — of this kind of economical gain.

The distinction between nominal and real definitions can be seen in a different light if definitions are looked at from the point of view of their utility. The characterization of nominal definition made in Cohen and Nagel (1961) — similar characterizations are made by other writers — that they are only stipulations concerning how verbal symbols are to be used, that they are not true nor false, that they cannot be premisses of any arguments, and that they do not extend our real knowledge (cf. Section 1), are not relevant for formal definitions in logic, as is easily seen from what we have learnt in this section and Section 3. It would be somewhat misleading to say, for example, that a formal sentence of elementary logic has no truth value, and false to say that it cannot be a premiss of any argument. The last-mentioned characteristic is not appropriate for nominal definitions — whether formal or not — in the sense in which Cohen and Nagel use this notion, for they say that all explicit definitions in mathematics are nominal, but, as we saw, we have reasons to assume that such definitions may extend our knowledge. Furthermore, the term 'nominal' is often used by Cohen and Nagel (1961) — and others, e.g., by Hempel (1952) and Robinson (1965) — about definitions whose stipulative character is only apparent or partial, their ultimate purpose being analysis or explication. For instance, the definitions (1.1) and (1.2) of Section 1 are syntactic representations of the intuitive idea of how the meaning of a connective is thought to be related to other connectives. There are stipulative elements in the definitions since there are different ways to construe the meaning of implication and since (1.2) introduces a notation, but, on the other hand, they are results of an analysis and explication of the notion of implication — and as such they extend our knowledge. It is evident that in fact most important definitions exhibit both features which are considered characteristic of nominal definitions and features considered characteristic of real definitions. Hence, it is questionable whether the distinction is relevant at all — and similar doubts are expressed by Cohen and Nagel themselves, as we saw in Section 1. The distinction could, perhaps, be defined so that it coincides with the distinction between definitions which are and definitions which are not fruitful, or on the basis of their complexity, or the like. A distinction of that kind is suggested in Hintikka and Tuomela (1970). By and large, however, the whole distinction seems to be unnecessary.

Department of Mathematical Sciences/Philosophy
University of Tampere

REFERENCES

Austin, J. L., 'Performative Utterances', in J. L. Austin, *Philosophical Papers* (ed. by J. O. Urmson and G. J. Warnock), Oxford University Press, Oxford, 1970, pp. 233–252.

Barwise, J., *Admissible Sets and Structures*, Springer-Verlag, Berlin, Heidelberg, New York, 1975.

Beth, E. W., 'On Padoa's Method in the Theory of Definition', *Indag. Math.* **15** (1953), 330–339.

Carnap, R., 'Testability and Meaning' *Philosophy of Science* **3** (1936), 420—468; **4** (1937), 1—40.

Chang, C. C., 'Some New Results in Definability', *Bull. Am. Math. Soc.* **70** (1964), 808—813.

Chang, C. C. and Keisler, H. J., *Model Theory*, North-Holland, Amsterdam, 1973.

Cohen, M. R. and Nagel, E., *An Introduction to Logic and Scientific Method*, Routledge & Kegan Paul Ltd, London, 1961.

Cresswell, M. J., *Logics and Languages*, Methuen, London, 1973.

Frege, G., *Die Grundlagen der Arithmetik*, Wilhelm Koebner, Breslau, 1884.

Goodman, N., 'The Problem of Counterfactual Conditionals', *Journal of Philosophy* **44** (1947), 113—128.

Hempel, C. G., *Fundamentals of Concept Formation in Empirical Science*, University of Chicago Press, Chicago, 1952.

Hintikka, J., 'Constituents and Finite Identifiability', *Journal of Philosophical Logic* **1** (1972), 45—52.

Hintikka, J., *Logic, Language Games and Information*, Clarendon Press, Oxford, 1973.

Hintikka, J. and Tuomela, R., 'Towards a General Theory of Auxiliary Concepts and Definability in First-Order Theories', in J. Hintikka and P. Suppes (eds.), *Information and Inference*, D. Reidel, Dordrecht, 1970, pp. 298—330.

Hintikka, J. and Rantala, V., 'Systematizing Definability Theory', in S. Kanger (ed.), *Proceedings of the Third Scandinavian Logic Symposium*, North-Holland, Amsterdam, 1975, pp. 40—62.

Kueker, K., 'Generalized Interpolation and Definability', *Annals of Mathematical Logic* **1** (1970), 423—468.

Lewis, D., *Counterfactuals*, Blackwell, Oxford, 1973.

Makkai, M., 'A Generalization of a Theorem of E. W. Beth', *Acta Math. Sci. Hungar.* **15** (1964), 227—235.

Makowski, J. A., Shelah, S., and Stavi, J., 'Δ-Logics and Generalized Quantifiers', *Annals of Mathematical Logic* **10** (1976), 155—192.

Pearce, D., *Roads to Commensurability*, D. Reidel, Dordrecht, 1977.

Pearce, D. and Rantala, V., 'Logical Aspects of Scientific Reduction', in P. Weingartner and J. Czermak (eds.), *Epistemology and Philosophy of Science*, Hölder-Pichler-Tempsky, Vienna, 1983.

Pearce, D. and Rantala, V., 'A Logical Study of the Correspondence Relation', *Journal of Philosophical Logic* **13** (1984), 47—84.

Przelecki, M., *The Logic of Empirical Theories*, Routledge and Kegan Paul, London, 1969.

Przelecki, M., 'On Identifiability in Extended Domains', in R.E. Butts and J. Hintikka (eds.), *Basic Problems in Methodology and Linguitics*, D. Reidel, Dordrecht, 1977, pp. 81—90.

Quine, W. V. O., *From a Logical Point of View*, Harper & Row, New York, 1953.

Rantala, V., *Aspects of Definability* (Acta Philosophica Fennica, vol. 29), North-Holland, 1977 [1977a].

Rantala, V., 'Prediction and Identifiability', in R. E. Butts and J. Hintikka (eds.), *Basic Problems in Methodology and Linguistics*, D. Reidel, Dordrecht, 1977, pp. 91—102 [1977b].

Rantala, V. and Tselishchev, V., 'Surface Information and Analyticity', in R. Bogdan (ed.), *Jaakko Hintikka*, D. Reidel, Dordrecht, 1987, pp. 77—90.
Robinson, R., *Definition*, Clarendon Press, Oxford, 1965.
Saarinen, E., 'Propositional Attitudes Are Not Attitudes towards Propositions', in I. Niiniluoto and E. Saarinen (eds.), *Intensional Logic: Theory and Applications* (Acta Philosophica Fennica, vol. 35), Helsinki, 1982, pp. 130—162.
Searle, J., *Expression and Meaning*, Cambridge University Press, Cambridge, 1979.
Suppe, F., 'The Search for Philosophic Understanding of Scientific Theories', in F. Suppe (ed.), *The Structure of Scientific Theories*, University of Illinois Press, Urbana, Chicago, London, 1977, pp. 1—232.
Suppes, P., *Introduction to Logic*, Van Nostrand, New York, 1957.
Svenonius, L., 'A Theorem of Permutation in Models', *Theoria* **25** (1959), 173—178.
Tarski, A., *Logic, Semantics, Metamathematics: Papers from 1923 to 1938* (translated by J. H. Woodger), Clarendon Press, Oxford, 1956.
Tuomela, R., *Theoretical Concepts*, Springer-Verlag, Vienna, New York, 1973.
Whitehead, A. N. and Russell, B., *Principia Mathematica to *56* (from *Principia Mathematica*, 1910), Cambridge University Press, Cambridge, 1973.

JAAKKO HINTIKKA

TOWARDS A GENERAL THEORY OF IDENTIFIABILITY

1. WHAT IS IDENTIFIABILITY?

What is identifiability, anyway, and what does it have to do with definitions and definability? The basic intuitive idea is clear. A concept (say, a one-place predicate P) occurring in a theory $T[P]$ is *definable* on the basis of this theory iff the theory determines the interpretation of P as soon as the interpretations of the other concepts occurring in $T[P]$ are fixed. More explicit, this is what, the definability of P on the basis of $T[P]$ means.

What is meant by determination here remains to be spelled out. Furthermore, there are of course other variants of the idea of definability, some of which will be characterized below.[1] These questions nevertheless need not detain us here.

It may happen, however, that the interpretation of P is not determined on the basis of the theory alone, but is determined by the theory *together with* a number of auxiliary *empirical results*, such as observations or experimental outcomes.

Then it is said that P is *identifiable* on the basis of $T[P]$. Such a notion of identifiability is unlike the notion of definability in that it is relative to a model M (of the language of $T[P]$), for the auxiliary empirical information has to pertain to some one "world", i.e., some model of the language of $T[P]$. (When I mention "models" in the rest of this paper, I shall mean models of the language of the underlying theory, unless something else is indicated.) We can make the notion of identifiability independent of M by so to speak integrating over the space of models, e.g., by stipulating that P is identifiable on the basis of $T[P]$ *simpliciter* iff it is identifiable in each model of $T[P]$.

The notion of identifiability is also relative to the class of possible observations, as it were to the totality of answers that nature can give to an inquirer's questions.

Throughout this paper, I shall use the definability or identifiability of a simple one-place predicate as an example. All the definitions and results are easily extended to other cases.

J.H. Fetzer, D. Shatz and G. Schlesinger (eds.), *Definitions and Definability: Philosophical Perspectives*, 161–183.

There are distinctions one has to make between different variants of identifiability. Some of them have parallels in the theory of definability as distinctions between different kinds of definability.

2. WHY DOES IDENTIFIABILITY MATTER?

Why is the notion of identifiability important? Several weighty reasons can readily be found.

First, the notion of identifiability has played a significant role in the actual methodology of certain empirical sciences. Indeed, it first caught theoreticians' eye in the context of actual theorizing in mathematical economics.[2] It is not hard to guess in the light of hindsight what the problem was. Suppose a concept (a parameter) like our predicate P occurs in a theory $T[P]$ which is intended for the actual mastery of empirical phenomena. For such a purpose, the theory $T[P]$ is useless unless the value of the parameter in question can be estabilished by means of auxiliary exmpirical information (observable data). Yet if the logical (mathematical) structure of $T[P]$ is of the right sort (or, from the vantage point of the intended use of the theory, of the wrong sort), this may be intrinsically impossible (impossible for logical and mathematical reasons). This is precisely what is meant by the nonidentifiability of the parameter in question. And this is precisely what happened with certain macro-economic theories. In order to be forewarned (and forearmed), econometricians have since developed methods of recognizing the conditions on which such nonidentifiability can or cannot occur, at least in the types of theories most frequently used in economics.

A simple example of an identifiability problem comes up when the underlying theory assumes the form of a pair of linear equations

$$\text{(1)} \qquad \begin{aligned} a_1x + b_1y &= 1 \\ a_2x + b_2y &= 1 \end{aligned}$$

if the determinant D

$$D = \begin{vmatrix} a_1 & b_1 \\ a_2 & b_2 \end{vmatrix}$$

is different from zero, the only observable values of the variables x and y are the solutions to the pair of equations (1). But, from such observations one can never determine the parameters a_1, b_1, a_2, b_2.

However, of $D = 0$ the two equations (1) do not have any solutions unless they are equivalent. Then the parameters a_1 and b_1 can be identified from observations concerning two pairs of values of the variables x_1, y_1, and x_2, y_2, i.e., as solutions of the following equations with respect to a_1 and b_1:

$$(2) \qquad \begin{aligned} a_1 x_1 + b_1 y_1 = 1 \\ a_1 x_2 + b_1 y_2 = 1 \end{aligned}$$

In spite of its simplicity, this example is characteristic of the conceptual structure of identification problems in econometrics.

I shall return below to the role of the concept of identifiability in other sciences. This concept belongs in any case squarely to the actual working methodology of several different actual empirical sciences.

3. IDENTIFIABILITY DISGUISED AS DEFINABILITY

Second, what is called in philosophical or methodological discussions "definability" frequently, perhaps typically, turns out to be identifiability and not definability in any of the several senses of definability studied by logicians. Hence the study of identifiability is an integral part of any comprehensive study of definability as the latter term is actually used by philosophers and methodologists.

In some cases, the assimilation of definability to identifiability has been made in so many words. For instance, Herbert Simon acknowledges the fact that he uses the term "definability" to all practical purposes as a synonym for (a generalization of econometricians' concept of) identifiability.[3]

There is nothing intrinsically wrong with such a usage. We shall in fact see that there are close ties between the notions of identifiability and definability. (*Vide* theorems 11 below.) However, there is little payoff in the employment of the notion of identifiability on a general philosophical or methodological level as long as there does not exist a general logical theory of identifiability. Such a theory was not available at the time Simon was writing. In a sense, my paper is one of the first serious attempts to develop a general theory of identifiability.

A comparison with definability is instructive here. Even though the majority of philosophers seem to be unaware of it, there exists a rich logical theory of definability which has a great deal of potential philosophical interest. This theory comprises among other things definitions of different types of definability, their interrelations, their criteria, etc. In contrast, there does not exist any equally explicit general theory of identifiability. Admittedly, it turns out that certain aspects of the general theory of definability are in fact highly relevant to the study of identifiability, too. This fact has its deep reasons in the logical structure of identifiability problems. But even these existing results have not been collected together and presented with anything like their proper motivation. One of the purposes of the present paper is to show a general logical theory of identifiability can be developed as an extension and as a partial analogue to the general logical theory of definability.

To return to the relation of identifiability and definability, another danger in using "definability" simply as another term for identifiability with unnecessary disagreements as to what definability "really is" or how scientists "actually use" the notion of definability. This risk is not merely an abstract possibility, either. The one-time protracted controversy between Herbert Simon and Patrick Suppes concerning the role of definitions in science turns out, in the light of 20/20 hindsight, to have been a classic case of a debate at cross-purposes.[4] Suppes was using the terminology of definitions and definability in its established logical sense, whereas Simon employed the same terminology to mean some sort of generalized notion of identifiability for which there did not exist at the time any exact theory (outside the special cases studied by the likes of econometricians).

Simon was undoubtedly right in one important respect, however. When the terms "defining", "definition", and "definability" were used in traditional foundational or methodological discussions, what was meant was usually identifiability, not definability. For instance, when Ernst Mach posed the problem as to whether the concept of mass is definable in Newtonian mechanics, he was not asking whether the fundamental equations of (a suitable formulation) Newtonian mechanics logically imply an explicit definition of mass in terms of the other priorities of Newtonian mechanics, nor was he inquiring into any other variety of definability of the concept of mass on the basis of the fundamental laws of classical mechanics.[5] What Mach was asking was whether these laws enable a physicist to determine in each and every case the value of a point mass by means of suitable observations. In other words, Mach's question concerned, not definability, but identifiability.

4. THE INTERROGATIVE MODEL OF INQUIRY

How can we develop an explicit theory of identifiability? As was indicated above, it is technically possible to discuss some problems of identifiability simply by using in a suitable way the logical techniques employed in the existing logical theory of the different kinds of definability. This procedure would make it very hard to see the rationale of the different conceptualizations that have to be used here, and it would not be helpful heuristically, either, in bring out partial analogies between identifiability and definability. A more natural framework is offered by the *interrogative model of inquiry* which I have developed.[6] It is calculated to be a general conceptual tool in different kinds of investigations into epistemology and philosophy of science.

The interrogative model can be formulated in the form of a game. This game is what game theorists could call a game against nature, in that it is a two-person game in which only one player ("the Inquirer") is active whereas the other ("Nature") comes in only (or mainly) as a source of answers to the Inquirer's questions. In this paper, it will be assumed that all the answers are true (in the model to which the inquiry pertains).

At each stage of the game, the Inquirer has a choice between two kinds of moves, (i) a logical move (also called an inference move or *tableau*-building move) and (ii) an interrogative move. The former consists essentially in the Inquirer's drawing a deductive inference from earlier results. The latter consists essentially in the Inquirer's putting a question to Nature. If Nature comes up with an answer, this answer can be used by the Inquirer as an additional premise.

In order to be able to formulate the rules form these moves more explicitly, we can decide to use as our bookkeeping method — or, perhaps rather, the Inquirer's bookkeeping method — a variant of E. W. Beth's method of semantical *tableaux*.[7] Then the logical moves are steps of *tableau* construction. In speaking of them, we can use all the usual vocabulary pertaining to Beth's *tableaux*, such as the left column, the right column, *subtableau*, closure, etc. Since the basic idea of the *tableau* technique is extremely simple, using this technique will enable the reader to appreciate the interrogative model even without much knowledge of logic. Very briefly, a *tableau* which begins with T in the left column and C in the right column is the record of a thought-experiment, viz. an attempt to think of T as true and C as false in the same "world" (model). If such an attempt is frustrated in all directions (i.e., if the *tableau* closes), C cannot fail to be true if T is, i.e., C follows logically from T.

Using the *tableau* technique of course means assuming that we are dealing with (applied) first-order languages. I shall make this assumption for the purposes of this paper. Since we are dealing with interpreted languages, we have to make a sharp distinction between real ("proper") names (of individuals in the domain $do(M)$ of the model M to which the inquiry pertains) and "dummy names" introduced by the logical instantiation rules.

An interrogative game begins with a *tableau* where the only entry in the left column is an initial premise T. (It can of course also be a set of premises.) The only entry in the right column initially is the conclusion C to be proved. (It can also be a set of conclusions whose disjunction is to be proved.)[8]

In this paper, it is normally assumed that T is true in the model in question.

5. RULES FOR LOGICAL MOVES

I shall nevertheless modify even the rules for logical moves somewhat. I shall first give the rules and then indicate the rationale for the modifications.

The *tableau*-forming rules can be summarized as follows:

What occurs in the left column of a *subtableau.*	What may be added to the left column of the same *subtableau.*
(L.&) $(S_1 \& S_2)$	S_1 and/or S_2
(L.E.) $(\exists x)\, S[x]$	$S[a]$ ("a" = a new dummy name)
(L.A.) $(\forall x)\, S[x]$	$S[b]$ (for any *dummy name* or *real name* "b" occurring in the *subtableau*); $S[y]$ for any variable y which occurs free in the subtableau.

(L. ~ ∨) ~ $(S_1 \vee S_2)$	~ S_1 and/or ~ S_2
(L. ~ E) ~ $(\exists x)\, S[x]$	~ $S[b]$ (b as in (L.A.))
(L. ~ A) ~ $(\forall x)\, S[x]$	~ $S[a]$ (a as in (L.E.))
(L. ∨) $(S_1 \vee S_2)$	S_1 or S_2 (split the construction into two further *subtableaux*, with S_1 and S_2 as the new members).
(L. ~ &) ~ $(S_1 \,\&\, S_2)$	~ S_1 or ~ S_2 (as in L. ∨))

(R.&), (R. ∨), (R.A), R. ~ &), (R. ~ ∨) and (R. ~ E) are simply duals of the correspond rules on the left.

Occurring in the right column of a *subtableau*.	Add to the right column of the *subtableau*.
(R.E.) $(\exists x)\, S[x]$	$S[b]$ (for any dummy or real name b occurring in the same *subtableau*); $S[y]$ for any variable y which occurs free in the *subtableau*.
(R. ~ A) ~ $(\forall x)\, S[x]$	~ $S[b]$ (b as in (R.E))

Closure rules:

(C.T) A *tableau* is closed iff all its *subtableaux* are closed.

(C.L) A *subtableau* is closed if S and ~ S occur in its left column (for any S).

(C.R) A *subtableau* is closed if S and ~ S occur in its right column (for any S).

(C.LR) A *subtableau* is closed if S occur in both of its columns (for any S).

By way of an explanation of these rules, and especially of their difference from the usual *tableau* rules, the following can be said:

The basic idea of the *tableau* method is to see if the purported conclusion C actually follows from T by assuming that T is true but C is false and then by reducing this counter-assumption *ad absurdum.* Hence everything in a left column is thought of as being true while everything in a right column is thought of as being false. The usual *tableau* rules can be based on this idea. But in an interrogative inquiry something else is also assumed. It is assumed that what is in a left column is *known* to be true. For this reason nothing can be moved from the right column to the left one. For this reason, no traffic from the right to the left can be allowed. Hence each column must unfold by itself.

6. RULES FOR INTERROGATIVE MOVES

Rules for question-answer moves make use of the notion of the *presupposition* of a question. This notion does not have to be analyzed here, however. In interrogative moves, all action takes place in the left column.

(Q.P) If a disjuction $(S_1 \vee S_2)$ occurs in the left column of a *(sub)tableau,* the Inquirer may use it as a presupposition of the propositional question

$$(S_1 \text{ or } S_2)?$$

The two possible answers are S_1 and S_2. If Nature gives one of the answers, it is added to the left column of the same *(sub)tableau.*

(Q.Wh) If $(\exists x)S[x]$ occurs in the left column of a *(sub)tableau,* the Inquirer may use it as the presupposition of the wh-question.

Who (what, where, when, . . .), say x, is such that $S[x]$?

or in general,

Which individual, call it x, is such that $S[x]$?

The possible answers are of the form

$$S[b]$$

where "b" is a name of a member b of the domain $do(M)$ of the model

M to which the inquiry pertains. If Nature gives an answer of this form, it is added to the left column of the same *(sub)tableau.*

Because the table construction may branch, it may happen that the presupposition of a legitimate question is false. Then no true answer is available. In such cases, Nature is allowed to give any possible answer (of the right form). The same holds if the same column contains an earlier entry which is not true. Otherwise, Nature's answer must be true.

Since dummy names do not have any specific bearer, none of Nature's answers is allowed to contain dummy names. The same goes for free variables.

7. GENERAL REMARKS ON INTERROGATIVE DERIVABILITY

If the Inquirer has managed to close the game *tableau*, it is said that she or he *has derived C from T interrogatively*. If the Inquirer has a strategy in the interrogative game by means of which she or he can win (close the *tableau*) no matter what Nature does, it is said that *C is interrogatively derivable from T*. In other words, C is interrogatively derivable from T is and only if the Inquirer has a strategy which wins against any strategy of Nature's.

It is in order to point out that this does not mean that the *tableau* construction (which includes all of Nature's admissible answers) comes to an end after a finite number of steps. Even if each branch closes, there may nonetheless be arbitrarily long branches, because Nature can sometimes choose her answers from an infinite set of possible answers.

The need of considering several *subtableaux* can be eliminated by replacing the branching rules (L.$\vee$) and (R.&) by the following:

(L.$\vee$.Nature) When $(S_1 \vee S_2)$ occurs in the left column, the Inquirer may ask Nature to add S_1 or S_2 to the left column.

(R.&.Nature) When $(S_1 \& S_2)$ occurs in the right column, the Inquirer may ask Nature to add S_1 or S_2 to the right column.

In both rules, Nature must comply. The choice of the formula to be added is up to Nature in both cases. It is not hard to see that this change does not alter the notion of interrogative derivability. The modified version turns out to be heuristically useful in many proofs.

That C can be derived interrogatively from T in M (by reference to a fixed set of restrictions on Nature's answers) will be expressed by

(1) $M: T \vdash C$

From the definitions of the interrogative games above it can be seen in what sense (1) is a generalization of the usual relation of logical consequence

(2) $T \vdash \mathrm{C}$

In an extreme case, the Inquirer asks no questions (and/or Nature answers none). Then (1) and (2) coincide. Otherwise, (2) entails (1) but not always *vice versa.*

The notion of interrogative derivability has considerable philosophical interest. It is fairly obvious that what is meant by "consequences" and even by "deductive consequences" in contemporary philosophy of science is better captured by interrogative derivability. For instance, in expounding the so-called hypothetico-deductive model of science philosophers do not always realize how little is actually derivable purely deductively from an abstract scientific theory. I suspect that what the proponents of the model have intended is really a hypothetico-interrogative model of a science rather than a hypothetico-deductive model.

The interrogative derivability of C in a model M of T is in an interesting way intermediate between *a logical implication* from T to C and the *truth of C in M.* If questions play no role, we have the former. If questioning is pursued under optimal conditions (in a certain specifiable sense), C is interrogatively derivable in M without any T iff C is true in M.

One of the ways in which interrogative derivability differs from deductive consequence (logical derivability) is that the interrogative counterparts to cut elimination and similar proof-theoretical results do not hold for it.[9] Accordingly, the following game rule will add to the purview of interrogative derivability:

(L.Taut.) At any stage of the game, the Inquirer may add any tautological disjunction

$$(S \vee \sim S)$$

to the left column.

The dual rule will be called (R. Cont.). It entitles the Inquirer to add any contradictory conjunction (S & $\sim S$) to the right column.

The more complex the sentence S is quantificationally (e.g., the more layers of unlike quantifiers it contains), the more extra power is yielded by an application of (L.Taut.).

As I have shown elsewhere, the force of an application of (L.Taut.) is to assume that the Inquirer is prepared to raise the question whether or not it is the case that S.[10] Hence applications of (L.Taut.) by the Inquirer amount to extending the range of the Inquirer's awareness.

This observation suggests that we can safely assume a number of special cases of (L.Taut.). If the Inquirer has already formulated S explicitly, he or she can scarcely be unaware of the question of its truth. This motivates assuming, as I shall do in the rest of this paper, that (L.Taut.) holds if the following condition is satisfied:

> S is a subsentence of a formula in the *subtableau* in question or a substitution-instance of a subformula of a formula in the *subtableau* with respect to names occurring in the *subtableau*.

The general theory of interrogative derivability can be developed as an extension of the usual proof theory (theory of deductive derivability) and also to some extent as an extension of the usual first-order model theory. Now one of the systematic mainstays of the metatheory of the usual (deductive first-order) logic is the Interpolation Theorem (IT) due to William Craig:[11]

INTERPOLATION THEOREM (IT). If $T \vdash C$ but neither $\vdash \sim T$ not $\vdash C$, then there is an interpolation formula I such that
(a) $T \vdash I$
(b) $I \vdash C$
(c) Each free nonlogical symbol of I occurs both in T and C.

I shall not reproduce a proof of IT here. It can be proved easily by means of induction on the length of the proof of $T \vdash C$.

The Interpolation Theorem holds also when T and C are (possibly infinite) sets of formulas.

8. THE FUNDAMENTALS OF DEFINABILITY

In order to see how the usual theory of definability can be extended to identifiability, we first have to review some of the most basic concepts and results concerning definability. I shall use the definability of a one-place predicate P as the basis of a first-order theory $T[P]$ as a test case. For simplicity, it will be assumed throughout this paper that $T[P]$ has only infinite models.

EXPLICIT DEFINABILITY

$$T[P] \vdash (\forall x)(Px \leftrightarrow D[x])$$

where (i) P does not occur in the *definiens* $D[x]$; (ii) x is the only individual variable in $D[x]$; (iii) $D[x]$ contains only nonlogical constants (concepts) occurring in $T[P]$.

PIECEWISE DEFINABILITY

$$T[P] \vdash \vee_{i=l}^{i=k}(\forall x)(Px \leftrightarrow D_i[x])$$

with the same restriction on each $D_i[x]$ as on $D[x]$ in explicit definability.

In many ways, piecewise definability is the normal kind of definability (on the basis of a theory). Explicit definability can be thought of as a special case, i.e., a "uniform piecewise definability". In a complete theory, explicit and piecewise definability coincide.

IMPLICIT DEFINABILITY

(a) *Weak version*: Assume that $T[P]$ and a model M of $T[P]$ are given. Then there is for each $b \in do(M)$ (= the domain of individuals of M) a formula $F[x]$ without P such that the following are true in M:

(i) $F[b]$

Either

(ii) $T \vdash (\forall x)(F[x] \supset Px)$

or

(ii) $T \vdash (\forall x)(F[x] \supset \sim Px)$

(b) *Strong version*:

$$(T[P] \;\&\; T[P']) \vdash (\forall x)(Px \leftrightarrow P'x)$$

where $T[P']$ is like $T[P]$ except for containing a new one-place predicate P' wherever $T[P]$ contained P.

Note: This use of the term "implicit definition" seems to go back to Beth. It differs from the most usual meaning of "implicitly defined" among philosophers, which assumes that a concept is "defined" already by the role it plays in a theory like our $T[P]$.

DEFINABILITY IN A MODEL. P is definable (on the basis of the other concepts of $T[P]$) in a model M of the theory $T[P]$ if and only if an explicit definition of P is true in M.

The main results concerning these metatheoretical concepts are the following:

(1) Assume that P is weakly implicitly definable in M on the basis of $T[P]$. Then it is definable in M if and only if there is a finite upper bound to the quantificational depth (maximal length of sequences of nested quantifiers) of the different formulas F in the definition of weak implicit definability.
(2) Strong implicit definability and explicit definability coincide (Beth's Theorem).[12]
(3) Definability of P in every model of $T[P]$ is equivalent with piecewise definability (Sevenonius).[13]

The most crucial of these results is (2). It can be proved easily with the help of the Interpolation Theorem.

Notice that Beth's Theorem is very handy when one has to try to find out whether a concept is definable on the basis of a theory. The usual techniques of deduction cannot be applied directly for the purpose of testing the applicability of the definition of explicit definability, for one does not normally know ahead of time what the definiens $D[x]$ is supposed to be. In contrast, in deciding questions of implicit definability one only needs to try to deduce $(\forall x)\ (Px \leftrightarrow P'x)$ from $(T[P]\ \&\ T[P'])$. A suitable proof of the IT shows how to construct the actual definition $D[x]$ from a successful deduction.

Moreover, if one does it by means of the *tableau* technique, one can use it so that the end product is either a proof of definability or else (at least in the limit) a model in which all the other concepts of $T[P]$ and of $T[P']$ have a definite interpretation but P and P' are interpreted differently. Proving nondefinability (sometimes also known as independence) by means of such a model is known as Padua's method.[14]

Beth's Theorem shows that Padua's method is in a sense the universal method of proving nondefinability in first-order theories, and that a Padua-type model can always be constructed by means of an attempt to establish (weak) definability.

9. EXTENDING THE INTERPOLATION THEOREM

The usual (deductive) Interpolation Theorem can be extended so as to become a tool in the study of identifiability. The extended theorem can be formulated as follows:[15]

EXTENDED INTERPOLATION THEOREM. Assume
(i) $M: T \vdash C$
(ii) Not $T \vdash C$
(iii) Not $M: T \vdash (S \,\&\, {\sim} S)$ (for any S)

Then there is a formula I (the interpolation formula) and individuals $b_1, b_2, \ldots, b_k \,\varepsilon\, do(M)$ such that
(a) $M: T \vdash I$
(b) $I \vdash C$
(c) Each predicate, dummy name, and free variable in I occurs in both T and C.
(d) Each (proper) name in I *either* occurs in both T and C *or* was introduced in the proof of (i) for the first time by an answer to a wh-question *or* was imported in the same proof from the right column to the left column by an application of (L.A) or (L. ~ E)
(e) No wh-questions are used in the proof of (a).
(f) All the answers used in the proof of (a) are used in the proof of (i).

An examination of the Extended Interpolation Theorem and its proof shows how the structure of the interpolation formula partially reflects the structure of the original proof of (i). The clause (d) above shows how the nonshared names of I reflect the applications of (Q.wh) and such applications of (L.A) as import a name to the left column from the right column. Furthermore, it can seen that the number of layers of quantifiers in I equals the number of applications of (L.A) or (L. ~ E) to dummy names or free variables occurring only in the right column plus the number of applications of (R.E) (or (R. ~ A)) to real or dummy names occurring only in the left column.

One remarkable feature of the Extended Interpolation Theorem is that it holds largely independently of what restrictions are imposed on Nature's answers, as long as they remain the same throughout the theorem. Admittedly, a few minor assumptions have to be made in the Extended Interpolation Theorem as to what questions will be answered by Nature. These assumptions are extremely natural, however, and will not be examined here in any detail.

10. DEFINING IDENTIFIABILITY

Now we have finally reached a point at which we can define different kinds of identifiability. The fundamental idea is clear on the basis of the preliminary explanations offered in sec. 1 above. In the same way as interrogative derivability is a generalization of deductive derivability, identifiability is a generalization of definability. Because of this analogy, the interrogative model of inquiry suggests a number of definitions which are roughly parallel with the corresponding definitions of different kinds of definability.

EXPLICIT IDENTIFIABILITY. *P* is explicitly identifiable in *M* on the basis of *T[P]* iff

$$(*) \qquad M\colon T[P] \vdash (\forall x)(Px \leftrightarrow D[x, a_1, a_2, \ldots, a_k])$$

where $a_1, a_2, \ldots, a_k \,\boldsymbol{\varepsilon}\, do(M)$ and where the *definiens* satisfies the same conditions as in explicit definability (except for the presence of a_1, a_2 $\ldots, a_k$).

Moreover, (*) must be established without the help of any answers which contain *P*.

PIECEWISE EXPLICIT IDENTIFIABILITY

$$M\colon T[P] \vdash \vee_i (\forall x)(Px) \leftrightarrow D_i[x, a_1, a_2, \ldots, a_k])$$

with the help of any answers which contain *P*. Each D_i must of course be like *D* in explicit identifiability.

STRONG IMPLICIT IDENTIFIABILITY

$$M\colon (T[P] \,\&\, T[P']) \vdash (\forall x)(Px \leftrightarrow \mathrm{P}'x)$$

without the help of any answers which contain *P* or *P′*

Here $T[P]$ and $T[P']$ are as in the definition of strong implicit definability.

WEAK IMPLICIT IDENTIFIABILITY. For each element b of $do(M)$, either

$$M: T[P] \vdash Pb \quad \text{or}$$
$$M: T[P] \vdash \sim Pb$$

without the help of answers containing P.

11. BASIC RESULTS

The basic results in the theory of identifiability include the following:

THEOREM 11.1 (Extended Beth's Theorem): Strong implicit identifiability implies explicit identifiability.

This can be proved by means of the Extended Interpolation Theorem essentially in the same way as its deductive (logical) counterpart.

Many of the same things as were said above of the role of Beth's Theorem in the theory of definability can be said of the Extended Beth's Theorem in the theory of identifiability. It opens up a way of testing by means of an interrogative argument whether a concept is (explicitly or strongly implicitly) identifiable in a model on the basis of a theory. (Cf. also sec. 14 below.)

Identifiability and definability differ from each other in that weak implicit identifiability does not imply strong implicit identifiability. There nevertheless are results relating the two.

THEOREM 11.2. Explicit identifiability implies strong implicit identifiability.

THEOREM 11.3. If P is explicitly or weakly implicitly identifiable in every model of $T[P]$, then it is explicitly definable on the basis of $T[P]$.

This theorem shows why it is so easy to confuse identifiability and definability with each other.

THEOREM 11.4. If P is piecewise explicitly identifiable in each model of $T[P]$, then

$$T[P] \vdash (\exists y_1)(\exists y_2)\ldots(\exists y_m) \vee_i (\forall x)(Px \leftrightarrow D[x, y_1, y_2, \ldots, y_m])$$

This theorem is closely related to a well-known result in the theory of definability.[16]

THEOREM 11.5. Assume that M: $T[P] \vdash Pb$ without answers containing P. Then there are $F[x, y_1, y_2, \ldots, y_k]$, and $a_1, a_2, \ldots, a_k \in do(M)$ such that

(i) M: $T[P] \vdash F[b, a_1, a_2, \ldots, a_k]$ (without answers containing P)

(ii) $T[P] \vdash (\forall x)(\forall y_1)(\forall y_2)\ldots(\forall y_k)(F[x_1, y_1, y_2, \ldots, y_k] \supset P_x)$.

(iii) There are no individual constants in $F[x, y_1, y_2, \ldots, y_k]$.

(iv) P does not occur in F.

Furthermore, the same holds for $\sim Pb$ instead of Pb.

Theorem 11.5 might be compared with Hempel's covering law model of explanation. It shows that there might be a little bit more to be said for the covering law model from a logical point of view than philosophers have recently acknowledged.

THEOREM 11.6. Assume that P is weakly implicitly identifiable in M on the basis of $T[P]$. Then it is strongly implicitly identifiabile if and only if there is a finite upper bound both to k and to the quantificational depth of F_2 in theorem 11.4.

These results will, I hope, give the reader the sense of what can be done in the general logical theory of identifiability.

There is a connection between the results stated here, especially the Extended Beth's Theorem, and the study of identifiability in econometrics. Assume that $T[P]$ is a set of equations. What the Extended Beth's Theorem says then is that P is identifiable on the basis of $T[P]$ iff this set of equations is solvable for P with respect to all the different available values of the other concepts of $T[P]$. Much of the econometric theory of identifiability can be thought of as spelling out the consequences of this observation.

12. IDENTIFIABILITY, MEASURABILITY AND OBSERVABILITY

An analysis of the concept of identifiability in terms of the interrogative model can be useful in philosophy, especially in the philosophy of science.

As a small example I can mention a frequently employed argument for the alleged incommensurability of classical and relativistic mechanics.[17] This argument for the alleged incompatibility of the two theories is based on the claim, which can be granted here, that certain concepts are identified in different ways in Newtonian and Einsteinian physics. That this kind of argument carries little weight is suggested by theorems 11.5 and 11.6. The former shows that the identification of P can take place in different ways for different individuals. Theorem 11.5 says, essentially, that the several identifying formulas can be combined in one.

A larger-scale application deals with the notions of observability and measurability. Philosophers of science have discussed extensively the contrast between observable concepts and theoretical concepts, but this discussion has not produced much genuine enlightenment.[18] The interrogative model shows that more distinctions have to be made here than are customarily made in the literature. The most important distinction that has to be made is between, on the one hand, what can be said of the answerability of questions containing a given concept and, on the other hand, the identifiability of a concept, in the different senses of identifiability.

The set A of available answers normally depends on the concepts these answers contain, but of course it normally depends on other factors as well, such as the quantificational depth of the answers.

It seems to me that the terms "observable" and "observability" are most naturally taken to pertain to the availability to answers to the Inquirer's questions to Nature (including one's perceptual environment). After all, observations can be thought of as a subclass of answers put to the observer's environment. Of course, finer distinctions between different variants of observability can be made here. In one sense of observability (observability$_2$), a concept is observable iff it occurs in an available answer to the Inquirer's question. In other words, observability$_2$ of all the concepts occurring in a potential answer is a necessary condition of its availability.

However, it can be required that all answers containing observable concepts be available, subject of course to possible other restrictions on the availability of answers. This variant will be called observability$_1$. It amounts to treating observability as a kind of sufficient condition of the availability of answers.

Both senses of observability are quite different from identifiability. As the notion of identifiability is defined here, it shares certain features with what is in the philosophy of science called measurability. Tentatively, and with certain important qualifications, identifiability could perhaps be considered a rational reconstruction of the idea of measurability.

In earlier discussions of observability and measurability, these notions are often assimilated in effect to one another. It is therefore important to heed the differences that obtain between them. For instance, observability (as I have defined it by reference to the interrogative model) is independent of any particular theory that the Inquirer may set up. In contrast, identifiability is a "theory-laden" notion. Not only is it defined so as to be relative to a theory; it is sensitive to the choice of the theory.

13. IDENTIFICATION AND UNCERTAINTY

These suggestions offer indications of the potential uses of the interrogative model and of the notion of identifiability that can be defined by its means. As a final whetter of the reader's appetite, I shall mention one more possibility for further development.[19]

Consider once again, a model M of a theory $T = T[P, Q]$, which is now assumed to contain two one-place predicates P and Q, whose identification is at issue. For some individuals $b \in do(M)$ one can establish

(1) $\quad M: T[P, Q] \vdash Pb$

without using either P or Q in Nature's answers. For others, one can likewise establish

(2) $\quad M: T[P, Q] \vdash \sim Pb$

For still others, neither (1) nor (2) can be established. The class of such individuals is called the uncertainty range of *P* or *UNC* (*M:P*). The uncertainty range of *Q* is defined likewise. The uncertainty range thus is, in effect, the range of entities in which a concept *P* cannot be identified. (More strictly, in the special case considered here, is something missing here? cannot be identified in terms of concepts other than *P* and Q.) As was suggested in the preceding section, we might also speak of the range of cases where a concept cannot be measured.

Now it can happen depending on what the theory *T*[*P, Q*] is like, that the two uncertainty ranges *UNC*(*M:P*) and *UNC*(*M:Q*) vary inversely when we move from one model *M* of the theory to another, say *M**. In other words, when one uncertainty range gets smaller, the other one gets larger. (How these relative sizes are measured need not concern us here.) The more accurately you can measure one concept, the more uncertainty there is with respect to the other.

This, of course, is precisely the phenomenon instantiated by Heisenbergian uncertainty in quantum theory, except that we are there dealing with the measurability of two time-dependent functions which assign to each particle a position and a momentum. Thus a theory of identification opens a new way of approaching of the problems connected with Heisenbergian uncertainty. For instance, we can see at once that the presence of such uncertainty can in principle be seen from the logico-mathematical form of a theory, just as nonidentifiability of a parameter in an econometric theory can be seen from the mathematical structure of the theory. (Cf. sec. 2 above.)

14. THE PLACE OF IDENTIFICATION IN INTERROGATIVE INQUIRY

From what has been said, it might appear that identification problems represent one particular application of the interrogative model of inquiry and of the techniques of inquiry that can be formulated by reference to this model. In reality, the theory of identifiability developed here is an integral and central part of the theory of interrogative inquiry itself.

In order to see in what sense this is the case, let us recall that the types of interrogative inquiry so far studied here are of a particular kind in that the inquirer's aim is to prove a given predetermined conclusion C. This is a very special and restricted typed of inquiry. It may occupy a privileged position in the pure theory of interrogative inquiry in that properties determine the logical properties of other kinds of inquiry. However, it is by far not the most interesting or important type of inquiry in practice. In the most interesting cases, the aim of an inquiry is to answer a question rather than to establish a conclusion already known. How can such types of inquiry be conceptualized?

The answer is fairly simple for propositional questions. For instance, to answer a yes-or-no question "B or not-B?" by means of an interrogative inquiry, the Inquirer can try to derive B interrogatively and independently try to derive $\sim B$ (in both cases from the same given theoretical premise T).

But what about wh-questions? For instance, suppose the question to be answered through interrogative inquiry is "Who are the P's?", in other words, "Who (or what) are the individuals x (members of the domain $do(M)$ of individuals of the model M in which the inquiry takes place) which satisfy 'Px'?" This question is obviously the same as: How can P be identified on the basis of T in M?

This question falls in two parts: Is P identifiable in M on the basis of T? If so, what is the identifying pseudo-definition of the form

$$(\forall x)(Px \leftrightarrow D[x, a_1, a_2, \ldots, a_k])$$

and what in particular is the *definiens* $D[x, a_1, a_2, \ldots, a_k]$?

This relates in a sense wh-questions and ways of trying to answer them to interrogative inquiry via the notion of identifiability. However, it does not yet enable the Inquirer to try to answer a wh-question, say "Who are all the P's?" by means of a single process of interrogative inquiry.

It is here that the Extended Interpolation Theorem and its consequence, Extended Beth's Theorem, become crucial ingredients in a general theory of interrogative inquiry. We have just seen that wh-questions are in the theory of interrogative derivability to be construed as questions of identifiability. Now the Extended Beth's Theorem shows how such questions can be answered by a single process of interrogative inquiry. In order to find out whether P is identifiable in M on the basis of $T[P]$ the Inquirer can try to derive

$$(*) \qquad (\forall x)(Px \leftrightarrow P'x)$$

interrogatively from $(T[P]\ \&\ T[P'])$ in M without the help of answers containing either P or P', where P' is a new one-place predicate. Moreover, if (*) is so derivable, from a closer examination of the proof of the Extended Interpolation Theorem it can be seen how the *definiens* can be read off from the interrogative argument that led to (*).

This, then, is how one can try to answer wh-questions by means of interrogative inquiry.[20] In order to answer a question like, "Who are the *P*'s?, the inquirer can try to derive (*) interrogatively from ($T[P]$ & $T[P']$) in the "world" *M* to which the inquiry pertains. If the attempt succeeds, from the resulting argument one can construct the *definiens* which yields the desired answer: the *P*'s are precisely those individuals which satisfy

$$D[x, a_1, a_2, \ldots, a_k]$$

This result shows convincingly the scope of the technique of interrogative inquiry.[21]

Department of Philosophy
Boston University

NOTES

[1] For a more complete treatment of the subject, see V. Rantala's *Aspects of definability*, Acta Philosophica Fennica, Vol. **29**, No. 2 (1977), North Holland, Amsterdam.

[2] See Cheng Hsiao, 'Identification', in Z. Griliches and M.D. Intriligator (eds.), *Handbook of Econometrics*, Vol. **1**, Ch. 4, North Holland, Amsterdam, 1983; F. Fisher, *The Identification Problem in Econometrics*, McGraw-Hill, New York, 1966.

[3] See H. A. Simon, 'The Axiomatization of Physical Theories', *Philosophy of Science*, Vol. **37**, No. 1 (1970), pp. 16—26.

[4] See H. A. Simon, *op. cit*; 'The Axioms of Newtonian Mechanics', *Philosophical Magazine*, Series 7, Vol. **33** (1947), pp. 888—905; 'The Axiomatization of Classical Mechanics', *Philosophy of Science*, Vol. **21** (1954), pp. 340—343; 'Definable Terms and Primitives in Axiom Systems', in L. Henkin, P. Suppes and A. Tarski (eds.), *The Axiomatic Method*, North-Holland, Amsterdam, 1959; J. C. C. McKinsey, A. C. Sugar and P. Suppes, 'Axiomatic Foundations of Classical Particle Mechanics', *Journal of Rational Mechanics and Analysis*, Vol. **2**, pp. 253—272; J. C. C. McKinsey and Patrick Suppes, 'Transformations of Systems of Classical Particle Mechanics', *ibid.*, pp. 273—289; P. Suppes, *Introduction to Logic*, Van Nostrand, Princeton, 1957, chapter 8; M. Jammer, *Concepts of Mass in Classical and Modern Physics*, Harvard U.P., Cambridge, 1961, especially ch. 9.

[5] E. Mach, *Die Mechanik in ihrer Entwicklung*, F. A. Brockhaus, Leipzig, 1883.

[6] See, e.g., J. Hintikka, 'The Logic of Science as Model-Oriented Logic', in P. Asquith and P. Kitcher (eds.), *PSA 1984*, Vol. **1**, Philosophy of Science Association, East Lansing, MI, 1984, pp. 177—185; 'Knowledge Representation and the Interrogative Approach to Inquiry', in M. Clay and K. Lehrer (eds.); *Knowledge and Skepticism*,

Westview Press, Boulder, CO., 1989; 'What Is the Logic of Experimental Inquiry?', *Synthese* vol. **74** (1988), pp. 173—188.

[7] See E. W. Beth, 'Semantic Entailment and Formal Derivability', *Mededelingen ven der Koninklijke Nederlandse Akademie van Wetenschappen*, Afdeling Letterkunde, N.R., Vol. **18**, No. 13 (1955), Amsterdam. This technique is formally speaking but a mirror image of a version of a Gentzen-type sequent calculus.

[8] There are of course other types of interrogative inquiry in which the aim of the game is not to prove a predetermined conclusion but to answer a question. I shall return to this matter in sec. 14 below.

[9] For these results and for their background see, e.g., J. Barwise (eds.), *Handbook of Mathematical Logic*, Part D, North Holland, Amsterdam, 1977.

[10] See my paper, 'Knowledge Representation and the Interrogative Model of Inquiry', in Marjorie Clay and Keith Lehrer (eds.), *Knowledge and Skepticism*, Westview Press, Boulder, Colorado, 1989, pp. 155—183.

[11] W. Craig, 'Three Uses of the Herbrand-Gentzen Theorem in Relating Model Theory to Proof Theory', *Journal of Symbolic Logic*, Vol. **22** (1957), pp. 269—285.

[12] E. W. Beth, 'On Padua's Method in the Theory of Definition', *Indagationes Mathematicae*, Vol. **15** (1953), pp. 330—339.

[13] L. Svenonius, 'A Theorem About Permutation in Models', *Theoria*, Vol. **25** (1959), pp. 173—178.

[14] E. W. Beth, *op. cit.*

[15] This theorem was first formulated in Jaakko Hintikka and Stephen Harris, 'On the Logic of Interrogative Inquiry', in A. Fine and J. Leplin (eds.), *PSA 1988*, Vol. **1**, Philosophy of Science Association, East Lansing, MI, 1988, pp. 233—240.

[16] See C. C. Chang, 'Some New Results in Definability', *Bulletin of the American Mathematic Society*, Vol. **70** (1964), pp. 808—813; M. Makkai, 'A Generalization of a Theorem of E. W. Beth', *Acta Math. Acad. Sci. Hungar.*, Vol. **15** (1964), pp. 227—235.

[17] For this issue, see, e.g., J. Giedymin, 'Logical Comparability and Conceptual Disparity between Newtonian and Relativistic Mechanics', *British Journal for the Philosophy of Science*, Vol. **24** (1973), pp. 270—276; David Pearce, *Roads to Commensurability*, D. Reidel, Dordrecht, 1987, pp. 154—158.

[18] See, e.g., R. L. Causey, 'Theory and Observation', in P. Asquith and H. Kyburg, Jr. (eds.), *Current Research in the Philosophy of Science*, Philosophy of Science Association, East Lansing, MI, 1979, pp. 187—206; R. Tuomela, *Theoretical Concepts*, Springer-Verlag, Wien & New York, 1973.

[19] See here J. Hintikka and P. Sibelius, 'Identification and Heisenbergian Uncertainty', forthcoming.

[20] Notice that in the principal wh-question to be answered through inquiry the wh-question is in effect given what I have called the universal reading, while the "small" questions through which it is answered is given an existential reading. (See here Jaakko Hintikka, *The Semantics of Questions and the Questions of Semantics, Acta Philosophica Fennica*, Vol. **28**, No. 4, Societas Philosophica Fennica, Helsinki, 1976, especially ch. 4.) This is as it ought to be. The Inquirer is trying to extract the maximal information from the questioning procedure, and hence prefers the universal reading of wh-questions. Nature is operating with the contrary purpose, and hence chooses the reading of wh-questions on which they have the least informative answers.

[21] In writing this paper, I have profited greatly from cooperation with Stephen Harris.

PART IV

EPISTEMIC DIMENSIONS

DAVID SHATZ

EPISTEMIC TERMS AND THE AIMS OF EPISTEMOLOGY

No area of contemporary philosophy has been so overrun by "definitions," "accounts," "analyses," and "explications" as has the theory of knowledge. Ever since Edmund Gettier published a three-page paper in 1963 which produced counterexamples to the "justified true belief" account of "*S* knows that *p*," philosophers have sought to formulate the "right" set of necessary and sufficient conditions not only for knowledge, which was Gettier's concern, but also for allied notions like justified belief and rational belief.[1] The result is a vast literature consisting largely of examples and counterexamples, and by now numbering close to 1200 items.

Looking over this literature, one feels prompted to express two kinds of concerns about it: methodological and evaluative. On the methodological end, one wonders about such questions as: How can extraordinarily complex conditions such as those proposed in the literature be invoked to *explain* the linguistic intuitions of ordinary, philosophically untutored people? How can one produce examples or counterexamples to show that "we" do or do not subscribe to this-or-that account of knowledge/justified belief, when one's interlocutor (who is part of the "we") insists that he or she subscribes to a different account? How could the interlocutor be using one analysis (the one "we" use) in forming intuitions, but an opposing one when called upon to articulate a theory? What shall we do about the fact our intuitions are so vague in many cases?

Besides these, we have basic methodological questions about how analyses, accounts, explications, and definitions should be defended: Does one test a proposed theory solely by examining intuitions about individual cases? Or, should the proposed theory be tested by its ability to also connect up in interesting ways to larger epistemological issues? Should we look for a narrow match with intuitions about individual cases, or seek some reflective equilibrium between intuitions about principles and intuitions about individual cases?

This last set of methodological issues gives rise to what I have called the evaluative question: Is the definition of epistemic terms an impor-

J.H. Fetzer, D. Shatz and G. Schlesinger (eds.), *Definitions and Definability: Philosophical Perspectives*, 187—202.

tant philosophical activity? On this issue opinions are divided. Some philosophers have argued that the project is not worthwhile or even historically significant.[2] In contrast, others have called upon their favored accounts to illuminate other areas of philosophy: the nature of inference, mathematical Platonism, linguistic nativism, the mind-body problem, theory construction in science, and, most importantly, philosophical skepticism.[3]

In this paper I should like to consider the last of these subjects, philosophical skepticism. Numerous writers, from different camps, have thought that the analysis of knowledge and justified belief is linked, one way or another, to the problem of skepticism. I shall consider three ways in which they have tried to draw the connection: the possibility argument; the ordinary believer argument; and the discovery argument. My thesis is that none of these arguments are successful, and that accounts of knowledge and justified belief do not have anything interesting to do with skepticism.

1. THE POSSIBILITY ARGUMENT

The connections I should like to consider between definitions and skepticism are afforded, or allegedly afforded, by a general approach known as externalism. Broadly speaking, externalism about justified belief says that the property that makes a belief justified is one of which the justified believer might be ignorant; and externalism about knowledge says that the property that makes a belief into an item of knowledge is one of which the knower might be ignorant. The most common version of externalism is "reliabilism" — the view that knowledge and/or justified belief is belief formed or sustained by reliable processes and methods. To be sure, certain theories antithetical to reliabilism might also qualify as externalist, for example, coherence theories of justified belief and "undefeated justified true belief" accounts of knowledge. Nevertheless, it will ease our discussion to consider externalist responses to skepticism in their reliabilist version.

The most common way of relating externalism to skepticism is via *the possibility argument*. Reliabilists oftens argue as follows. Skeptics claim that we do not know what we think we know and are not justified in believing what we think we are justified in believing. They show this by showing that all dialectical arguments for common sense fail. Now, if, to know that *p* or be justified in believing that *p*, we need an argument that will be dialectically effective against the skeptic, then, alas, the skeptic is right; for the history of epistemological argumentation shows clearly that we cannot provide a proof for our commonsense views. But suppose that knowledge and/or justified belief is defined as belief formed or sustained by reliable processes. We then have knowledge or justified belief just in case our beliefs are, *in point of fact*, formed or sustained by reliable processes. Skeptics have no proof that our beliefs are *not* formed by reliable processes. So skeptics have no right to assert that we do *not* have knowledge or justified belief. Possibly, we do have knowledge and justified belief.[4]

Philosophers have heaped a great deal of criticism on reliabilist accounts, and I quite agree that these accounts run into serious counterexamples. On reliabilist accounts, a clairvoyant has knowledge and justified belief even if he has no reason to trust his beliefs, indeed, even if he has reason to distrust them. Isn't this counterintuitive?[5] Perhaps, but it is not easy to refute reliabilism in this way. Reliabilists may simply refuse to acknowledge the force of the counterexamples, and it just isn't clear that antireliabilists have a way of converting them to their side. That reliabilist definitions seem counterintuitive *to antireliabilists* is no reason for *reliabilists* to abandon their definitions. In addition, it can be shown that non-reliabilist accounts also face such problems. That is, we can construct cases in which *S* satisfies a certain condition alleged to constitute a sufficient condition of justified belief, but isn't aware that he does, or believes that he doesn't; for many theories of justification, including non-reliabilist ones, we can show that intuitively *S* does not have justified belief in such cases. Moreover, some reliabilist accounts, notably Alvin Goldman's most recent one, impose conditions which block the clairvoyance counterexamples and more generally seek to accommodate some of the antireliabilist's intuitions.[6] Finally, and most importantly for our purposes, reliabilists could argue that precisely because their definitions turn back the skeptical challenge, they ought to be adopted. After all, theories of what knowledge is or what justified belief is should at least partially be dictated by the aims of epistemology, rather than by case-by-case intuitions alone; and one aim of epistemology is vanquishing the skeptic. Immunity from skepticism may represent so significant a gain that it constitutes a reason for embracing reliabilism as against other accounts of epistemic terms. In light of the limitations of refutation by counterexample, perhaps the most antireliabilists can hope for is this: to show reliabilists that their accounts have no payoff in other epistemological issues, no impact on the skeptical challenge.

What is the skeptical challenge? The reliabilist response clearly takes the skeptic to be arguing for one form of skepticism, *epistemic* skepticism (ES):

ES: We do not have knowledge or justified belief.

The skeptic argues for ES on the basis of *dialectical* skepticism (DS):

DS: There is no dialectically effective argument
for our first-level common-sense beliefs or our
second-level common-sense beliefs.

But DS does not support ES, say reliabilists, because neither knowledge nor justified belief requires possession of a dialectically effective argument. Hence the skeptic's argument is a *non sequitur.*

What we really have in reliabilism's approach to skepticism, then, is a consistency argument: the thesis that we have knowledge and justified belief (i.e., the negation of skeptical thesis ES) is consistent with DS. Knowledge and justified belief are *possible.*

Against this reliabilist response I would urge that what is really important about the skeptical challenge is DS, not ES. To show this, I present two parodies of the possibility argument.

Parody 1: A philosopher defines knowledge as true belief, period. He then goes on to argue that, with his account, our beliefs would constitute knowledge *if* they are true. We respond that such a reply to the skeptic is of no real help; obviously we realized our beliefs *might* be true. Why should explicating knowledge as reliable true belief serve any better with respect to the skeptic's challenge? We have no better an argument for the reliability of the process by which the belief was formed, than we do for the truth of the belief.

Parody 2: Assume a reliabilist account of epistemic terms. Ed believes that there are no physical objects. Rather, in Ed's view, our various sensory experiences are fed to us by agents — scientists, God, or what-have-you. Whenever Ed has "cat-like" experiences, he hypothesizes: God (or the scientists) are now stimulating me to have "cat-like" experiences. We of course are "skeptical" of Ed's hypothesis; in fact, we deny it. But suppose we object to Ed that he lacks enough warrant for his hypothesis. In a manner isomorphic to the reliabilist's reply to the skeptic, Ed could fend off "skeptical" objections to his "agency" hypothesis. "That belief *may* be reliable," Ed will say. In other words: if I really am being stimulated by agent x, then I know that I am now being stimulated by *x*, because I then satisfy the reliabilist's conditions for knowledge. Consider, for example, what we get when we apply Robert Nozick's crucial condition: I know that I am now being stimulated by *x* to have cat-like experiences provided that, if *x* weren't stimulating me to have such experiences, I wouldn't be having the experiences. This subjunctive conditional is true in the circumstances Ed envisages. And for that matter, we cannot be "skeptical" of any skeptical hypothesis Joe might endorse.[7]

Parody 2 brings out an important constraint on "solutions" to the skeptical problem. Any reply to a skeptic who puts common sense in doubt should make clear why those who endorse common-sense hypotheses are in a *better* epistemic position than those who endorse skeptical ones. The possibility argument of the reliabilist does not meet this demand. The problem, notice, is perfectly general; it plays off the notorious "underdetermination" of common-sense hypotheses. If someone endorses a hypothesis which is incompatible with the common-sense one (that physical objects produce veridical experiences in us), but which is equally compatible with the data, he could claim to turn back any "skepticism" about his hypothesis when that skepticism is predicated on the absence of a good enough dialectical argument for his view. True, given *our* metaphysics, his beliefs will not be reliable (or even true). But the fact remains that, on *his* metaphysics, his belief is true and reliable. And we have no dialectically effective argument against his metaphysics.

What these parodies show is that skepticism is really a matter of DS, not ES. If skeptics ever framed their view as ES, this was because skeptics had few of our post-Gettier worries about defining epistemic terms, blithely assumed that knowing or being justified entails having a dialectically effective argument, and adopted ES as a *facon de parler*. Had skeptics realized that reliabilist accounts were available which would break the nexus between DS and ES, they never would have formulated their point in terms of ES. In sum, the possibility argument leaves DS standing; and DS, not ES, represents the real core of the skeptical challenge.

In fairness to externalists, it should be noted that, for some externalists, externalism salvages only the *coherence* or *consistency* of our common-sense framework, and does not turn back skepticism itself. That is, externalism helps explain how on the one hand we attribute knowledge (and justified belief) while on the other we recognize the unanswerability of skeptical challenges. We do so, the externalist explains, because (1) *we* do not take the unanswerability of skepticism to entail lack of knowledge or justified belief, and (2) *we* believe that our beliefs *are* reliable. Our practice of attributing knowledge and justified belief is therefore coherent, and is not undermined by our concurrent belief that we lack a dialectically effective argument for common sense.[8] I do not wish to reject this claim of coherence. But this is a far cry from *answering* skepticism in a stronger form.

2. THE "ORDINARY BELIEVER" ARGUMENT

Finding an account of knowledge or justified belief on which knowledge or justified belief does not require a dialectically effective argument will not *suffice* to refute skepticism. The reason, again, is that the crucial problem of skepticism is DS, not ES. However, one might still want to boast a theoretical advantage for externalist accounts vis-à-vis skepticism. Granted that showing ES to be unproven does not help reply to DS; still, one might say, providing a dialectically effective argument for common sense would not suffice to solve *all* skeptical problems. For ES, it will be argued, would not be turned backed by possession of a dialectically effective argument; only reliabilism affords a way of replying to ES. So, the claim will be not that reliabilism is *sufficient* to turn back all skeptical challenges — reliabilism is, after all, impotent against DS — but that it is *necessary* if we are to turn back ES.

To clarify this line of thought, let us consider an argument formulated by Elliot Sober.[9] Discussing Carnap's program of reconstruction, Sober writes:

> Carnap's project differed from that of cognitive psychology . . . in that he was not interested in how people *in fact* construct their physical object concepts and beliefs from an autopsychological basis. Rather, Carnap wanted to describe how people *might* achieve this advance.

But, Sober then argues, Carnap's reconstruction is also meant to show why our beliefs are *rational* — the construction was to simultaneously be a justification." Here, unfortunately, "Carnap's retreat from psychological actuality makes it impossible for him to successfully discharge this normative task." Why? Because even if beliefs about the physical world are related to beliefs about mental states by Carnapian justification relations,

> this does not show that any of us is justified in believing in an external world. Maybe we believe this for entirely different reasons, or for no reason at all. Carnap's justificatory strategy *at most* shows that if we *in fact* had arrived at our beliefs the way he describes, then our beliefs *would be* justified . . . if . . . we are interested in whether we in fact do know anything, we must go beyond speculative reconstruction and see how beliefs are in fact arrived at . . . To refute scepticism, one must have a formulation of one rational constructive method, a characterization of the method that is psychologically real, *and* [my emphasis] an argument that the epistemologically sanctioned method and the psychologically real one bear an appropriate equivalence relation to each other . . . the question the sceptic forces us to face is: were our beliefs chosen in accordance with this policy [the epistemologically sanctioned one] . . . . Only if the answer is 'yes' can the sceptic be refuted. (Sober, 188/89)

According to Sober, skepticism cannot be refuted by dialectical arguments unless those arguments are psychologically real, i.e. attributable to real people.

Anthony Quinton expresses the same thought:

> If a type of belief is rationally reconstructed as depending on a kind of support which it never psychologically or empirically possesses, the validity of the reconstruction has the absurd result that no belief of that kind is ever in fact justified.[10]

The claim, then, is that a dialectically effective argument is not enough to meet the skeptical challenge. For skepticism can be answered only by showing that *real people* have knowledge and justified belief.

Sober and Quinton both endorse the following principle:

> *Attributability Constraint (AC)*: A line of reasoning *R* will not justify *S* in a belief unless *R* is psychologically attributable to *S*, i.e. it reflects a line of reasoning by which *S* arrived at the belief or which sustains the belief.

But where does their argument go from there? It is hard to see how AC, by its lonesome, generates any absurd result. For, if some philosopher really arrives at a correct rational reconstruction, the result cannot possibly be that "no belief of that kind is ever justified." Things are not *this* gloomy. For at least the philosopher who is advancing the justifying argument will be justified in the belief; although he or she would hold the belief independently of the argument's being available (human nature might be responsible for that), he or she can acquire justification through producing the argument. (A person can be justified in a belief even if there are multiple sufficient conditions in existence for his or her holding the belief.) Not only the philosopher, moreover, but the philosopher's audience of readers or listeners as well, will have the justification attributable to them. Hence, without some supplementary principles, it is hard to see what absurdity lies in the assumption that a good dialectically effective argument might be beyond the *ken* of "*ordinary* people."

I believe we can flesh out the Sober-Quinton position by adding the following constraint:

> *Ordinary Believer Constraint # 1 (OBC-1)*: A condition of adequacy for an approach to refuting skepticism is that, on the approach, *ordinary* believers (those not familiar with philosophical arguments) are justified in their beliefs.

From AC and OBC-1, it follows that a successful antiskeptical argument will have to be isomorphic to psychological reasoning processes of "*ordinary* people," and not merely the reasoning processes of some rare "real" people.

With a moment's reflection we can see why dialectical arguments that philosophers typically offer as replies to skeptics will not be attributable to ordinary believers. Consider an argument used by some champions of coherence theories of justification. According to these philosophers, a paradigm of a justified coherent system is the system that includes our best scientific theories about how perceptual mechanisms, memory, and introspection work. We start with the premise that spontaneous visual beliefs are justified, and move from there to a theory of perceptual processes; then the theory in turn justifies our believing the data we started with. Similarly, we might initiate a theory of memory by trusting our memory about certain data; those data may eventually be used to support a theory of memory.

I am not concerned now with just why the circularity of such arguments might not be vicious. Rather, let us suppose that the circularity is not vicious; further, let us suppose that scientific arguments of this sort are the only acceptable arguments on behalf of common sense.[11] Since most people are unaware of the scientific data and/or the inference pattern, most people would then not satisfy AC with respect to the coherentist's line of reasoning, and consequently would not hold justified perceptual beliefs. Granted, those familiar with the scientific pattern and conversant with the dialectical argument would be justified;

but that is not enough to satisfy OBC-1. This point can be extended even to versions of coherentism that do not construct a justification along scientific lines but instead employ a sophisticated version of common sense. Such arguments, too, are not readily attributable.[12]

There are other examples of justificatory patterns that philosophers regard as promising, but which OBC-1 excludes from serving as a reply to skepticism. Consider a view that justifies belief in an external world by "inference to the best explanation" from the way things appear. The concept of appearances, many have argued, is acquired *after* the concept of a physical object; hence the justificatory pattern in question would not reflect psychological reality. To take another example, a justificatory argument for common sense would have to rule out certain alternative hypotheses: imitations, illusions, brains-in-vats. "Ordinary" people do not rule these hypotheses out; they ignore them entirely unless there are contrary indications. The conjunction of OBC-1 and AC would make such arguments impotent in meeting the skeptical challenge of ES.[13]

So, the most promising candidates for "dialectically effective anti-skeptical arguments" are not attributable to ordinary believers. But now it seems fair to conclude that, if we accept OBC-1, the skeptic wins. What is Sober's reply?

Sober's reply would be that rules of correct reasoning are psychologically real, i.e. attributable to ordinary people. He argues for this thesis empirically.[14] But his move will not do. For even if our states of mind exhibited transitions that reflected perfectly the order of epistemological priority embodied in one correct rational reconstruction, more might be required to make the beliefs formed through that sequence *justified*. For instance, the reasoner might have to *appreciate* the epistemic link between one belief and another. It is one thing to move from autopsychological beliefs to physical object beliefs, from beliefs about past uniformities to beliefs about future events, from beliefs about external behavior to beliefs about other minds; it is another thing to understand *why* these transitions are licensed. Philosophers will try to supply these links (e.g., by arguments by analogy in the case of other minds, considerations about "best explanation" in that or other cases, etc.). There is no reason to think that these deeper, more reflective considerations will be psychologically real, i.e., instantiated by actual psychological states of the ordinary person. In this connection, it should be noted that, in order to do justice to experimental findings that attribute fairly widespread errors in reasoning to human beings, Sober attributes correct norms to ordinary people only at the level of *competence*, not at the level of *performance*. But it could be argued that AC is satisfied (i.e., a line of reasoning is attributable) only when performance matches competence. In short, the prospects of a theory satisfying OBC-1 without an incredible upsurge in philosophical education are, to say the least, dim.

At this juncture, reliabilism comes to the rescue. Suppose we have some dialectically effective argument for common sense. As noted, it is unlikely that such an argument will be attributable to "ordinary believers." What reliabilism says is that, for ordinary believers to be justified in their common-sense beliefs, they need not have *any* justificatory argument of the type philosophers propose. This is too strong a requirement on justified belief. Once we see what justified belief really requires, our reliabilist will argue, we will see that psychological realities are such that ordinary believers do hold justified common-sense beliefs. This "ordinary believer" approach is nicely summarized (albeit in a noncommittal tone) by Jaegwon Kim: ". . . it is expected to turn out that according to the criteria of justified belief we come to accept, we know, or are justified in believing, pretty much what we reflectively think we know or are entitled to believe."[15] We can define what Kim is alluding to by formulating the following constraint:

> *Ordinary Believer Constraint # 2 (OBC-2)*: A condition of adequacy for an analysis of justified belief is that the theory has the result that ordinary believers are justified in holding those beliefs that common sense holds them to be justified in believing.

Reliabilism demands of people no sophisticated reasoning to achieve justification. It doesn't demand beliefs about appearances, it doesn't demand the explicit ruling out of all alternative hypotheses, and so forth. Once committed to OBC-2, no one will want to impose conditions on justified belief any stricter than the criteria people actually meet. Reliable belief seems to be the strictest criterion people actually meet, so reliabilism becomes the front-running account of justified belief — on pain of succumbing to ES. It is this claim, that OBC-2 constrains definitions of epistemic terms, that I want to consider and reject.

To begin with, OBC-2 begs all questions against epistemological skeptics of the ES type: we must have knowledge or justified belief because we think we have it. The victory comes too cheaply. Skeptics do not begin by constructing some crazy account of knowledge or justified belief. Rather, they bid us to look closely at standards we assume in ordinary contexts, or at plausible principles of justified belief (e.g., that being justified requires having reasons), and then show us that by those very standards we *don't* have knowledge or justified belief. The OBC-2 based approach "succeeds" against the skeptic *only* by rigging the deck at the start. How can the criterion of justified belief save us from the skeptic, if the criterion was *motivated* and *argued for* by the consideration that, if adopted, it would save us from the skeptic?

Second, whereas the approach uses OBC-2 to support reliabilism, there is in truth a tension between OBC-2 and reliabilism. Suppose that cognitive psychologists are studying beliefs that we intuitively classify as justified. The psychologists discover, to their great surprise, that people arrive at these beliefs by a process whose reliability falls below the reliabilist's threshold for justified belief. Given a reliabilist theory, such beliefs would have to be reclassified as unjustified, since they are not reliably formed. But consider what OBC-2 tells us about this scenario. OBC-2 would constrain us to *redefine* what justified belief is, so as to conform to the fact that we regard these ordinary beliefs as justified. It would therefore compel us to dismiss reliabilism. Once we accept OBC-2, we can never wind up concluding that ordinary believers are not justified in their beliefs! In the face of the psychologists' discoveries, it would seem reasonable to say that the *concept* of justification is such that justified beliefs must still be explicated as beliefs that are formed by reliable processes. OBC-2, though, would be dropped.

Consider an analogy. Ethicists sometimes call upon us to revise our intuitions about the moral standing of certain creatures (nonhuman animals, adults in irreversible comas, infants, fetuses) by appealing to criteria of moral standing. Originally, these criteria were formed by (a) accepting the intuition that normal adult human beings have moral standing; (b) reflecting on possible explanations as to why normal human beings have moral standing; (c) embracing or eliminating certain possibilities (e.g., belonging to a particular species; possesssing a capacity for abstract reasoning; possessing a capacity to suffer) on the grounds that they conflict with ordinary intuitions about cases, or else on the grounds that their relevance/irrelevance is just *obvious*. Plainly, the fact that we *wind up* rejecting our earlier intuitions about the extension of "morally permissible" shows that ordinary intuitions are not sacrosanct.

Imagine the following reply by a proponent of OBC-2: "Look, you're right that OBC-2 is too strong in making conformity to ordinary intuitions a *condition of adequacy* for a theory of justified belief. Yes: we've got to give skepticism a fair shake. Yes: if reliabilism sets down for us the *correct* standard of justified belief, it follows that, if ordinary beliefs are not reliably formed, the claim that they are justified should be given up. Since OBC-2 precludes this kind of argumentation, it can't be right as it stands. And yes: the analogy to ethics shows that people can have wrong antecedent intuitions about the extension of certain terms.

"Nevertheless, you must realize that any theory of justified belief must begin by collecting cases of 'intuitively' justified belief. There is a strong prima facie case against any theory of justification that says that most people are not justified in most of their beliefs. Once a theory is constructed from these cases, our initial judgments about the data may have to be revised so as to strike a reflective equilibrium. But we should be *wary* of accepting any theory that denies the correctness of our ordinary intuitions about which beliefs are justified for whom. Furthermore, if we didn't accept OBC-2 in *some* form, we would not know where to look for initial data; we could not identify instances of the concept we are trying to explicate."

There are several things wrong with this reply. To begin with, "justified" is not a term of common discourse. Whereas in the case of 'know,' we have before us a term that enjoys currency among "ordinary" people, the same cannot be said of "is justified." Given that the very existence of a data base for accounts of "is justified" is in question, OBC-2 stands on shaky ground even in its weaker role as a condition of initial plausibility. Here, then, is another reason for rejecting OBC-2 in any form: "common sense" does not "hold people to be justified" in their beliefs. "Justified," at least in the sense philosophers are concerned with, is not a term of the common people at all. It is a term of the philosopher's art.

In addition, the reply that OBC states a condition for initial plausibility rather than a condition of adequacy does not take sufficient notice of the dialectical context in which non-reliabilist theories of justification are developed. Let us take coherentism as an example of a theory of justification that requires, for *S* to be justified in believing that *p*, some pattern of mental processes, methods, lines of reasoning, etc., which may not be attributable to ordinary believers. Coherentists, too, might like to have a theory that does not deprive ordinary people of justified beliefs. But they think that the problems plaguing those theories that meet this condition are so numerous and grave that we must despair of arriving at a good account that will satisfy even the weakened ordinary believer constraint. Typically, coherentists arrive at coherentism by reflecting on the concept of justification. By reflection they conclude that a believer must have access to those features of his belief that justify it; that the notion of a self-justified belief is unacceptable; and so forth. In the course of developing these theses coherentists will often appeal to examples designed to show that rival theories are too weak. Thus, they make heavy appeal to intuitions. These intuitions, however, are intuitions about whether person *S* is justified in believing that *p* in hypothetical case *C*. Intuitions of this type can be used to construct a theory that might, given AC, undercut or challenge intuitions about the extent to which ordinary believers hold justified beliefs. In short, coherentists are prepared to admit that, other things being equal, a theory that allows us to conclude that ordinary believers are justified in their beliefs is better than one that does not. But other things are not equal.

It is unlikely that there is one single argument which is *both* dialectically effective against skepticism *and* attributable to ordinary believers. Faced with this fact and assuming AC, we must either (1) conclude that ordinary believers are not justified in believing the things they believe and do not know the things they think they know, or (2) design an account of knowledge or justified belief that ensures that ordinary believers have knowledge and justified belief. I have argued that option (1) is better motivated than option (2).

3. THE DISCOVERY ARGUMENT

In closing out our survey of alleged links between skepticism and epistemic terms, I should like to consider the view that the correctness of skepticism is an *empirical* issue which presupposes a certain *conceptual* thesis. First let us hear Marshall Swain:

> I do not take it to be the business of an analysis of knowledge or justification to settle skeptical questions. The analysis of knowledge and justification is, however, relevant to skeptical challenges, for the skeptic and the nonskeptic must agree antecedently about tbe nature of knowledge and justification if they are to argue meaningfully whether there are any instances of either.[16]

Likewise, Alvin Goldman writes early on in his lengthy book, *Epistemology and Cognition*

> Any adequate response to skepticism must involve, or presuppose, analyses (or 'accounts') of key epistemic terms like 'knowledge' and 'justification.'[17]

For these writers, we develop an account of knowledge or justified belief solely by reflecting on intuitions; the analysis itself is "neutral" on skepticism. However, once armed with the account, we *discover empirically* whether skepticism in correct by inquiring into whether human beings satisfy the conditions set down in the account. Both Swain and Goldman are, in spite of differences, reliabilists. For them, therefore, the key question we must answer before answering "do humans have knowledge and justified belief?" is "are the processes by which human beings form beliefs reliable?"

There are two steps in the Goldman-Swain approach: conceptual analysis and empirical discovery. There are problems with each step. First, it is doubtful that invoking "intuitions" to justify a given account of knowledge or justified belief will serve any purpose when one's interlocutor is a skeptic. Skeptics will not agree with nonskeptics' intuitions that Joe or Marsha knows that *p*/is justified in believing that *p* in case *X*. To skeptics, "ordinary" intuitions are symptoms of inexact, nonreflective thinking. Thus it will be hard to elicit agreement between skeptics and nonskeptics over any proposed account of knowledge or justified belief; a "neutral" analysis may be difficult to obtain. Second, any attempt to determine the scope of knowledge or justified belief *empirically* by determining *empirically* whether ordinary people satisfy the conditions of a particular account will have to presuppose the reliability of empirical procedures. But skeptics (of the dialectical variety) challenge the reliability of empirical procedures. Consequently, any attempt to exhibit, *pace* skeptics, that we have knowledge and justified belief, will have to presuppose something that skeptics question, namely, the reliability of empirical procedures. The enterprise of testing the truth of skepticism empirically is pointless.

4. CONCLUSION

The upshot is this. (1) As against the possibility argument, defining knowledge or justified belief in an externalist way will not remove the skeptical challenge, because skepticism is a dialectical problem. (2) As against the ordinary believer argument, there is no good motive for insisting on conditions of knowledge that will ensure knowledge and justified belief for ordinary believers. (3) As against the discovery argument, we will neither be able to find an analysis of epistemic terms that both skeptics and nonskeptics will agree on, nor establish empirically, in a mutually agreeable way, whether we have knowledge or justified belief.

The best we can do is show that *assuming* the belief-forming methods we use are reliable (as we suppose they are), most people hold true beliefs formed in reliable fashion. But this is only to demonstrate the coherence of our common-sense framework; it does not show its truth. We have a set of epistemic procedures *we* think are reliable, and using *them*, we "discover" we have reliable belief. But no skeptic will be persuaded thereby that we have reliable belief.

In fact, a further problem lurks here. For could we really discover that we *don't* have knowledge and justified belief? To reliabilists, this amounts to the following question: can we really discover that human belief-forming processes are *not* reliable? Obviously, even to make *this* discovery (the discovery that cognitive processes are not reliable), we would have to suppose that the procedures we used to discover that fact can be trusted. How this assumption could be justified in the face of skeptical challenges is never explained by reliabilists. Our empirical inquiry into whether skeptics are right can proceed only if we ignore the skeptic's main concern, namely, that our procedures do not generate truth. And so, in sum, definitions of epistemic terms cannot generate a response to skepticism any more than concern over skepticism can generate an argument for accepting a particular definition.

Department of Philosophy
Yeshiva University

NOTES

[1] Gettier, E., 'Is Justified True Belief Knowledge?,' *Analysis* **23** (1963): 121–23.

[2] A particularly forceful critique is Mark Kaplan, 'It's Not What You Know That Counts,' *The Journal of Philosophy* **82** (1985): 350–63. See also Isaac Levi, *The Enterprise of Knowledge* (Cambridge, Mass.: M.I.T. Press, 1980), pp. 29–30; George Schlesinger, 'Do We Have to Know Why We Are Justified in Our Beliefs?,' *Mind* **89** (1980): 370–90, Esp. pp. 373–75.

[3] On skepticism, see especially: Fred Dretske, 'Conclusive Reasons,' *Australasian Journal of Philosophy* **49** (1971): 1–22; Alvin I. Goldman, *Epistemology and Cogni-*

tion (Cambridge, Mass.: Harvard University Press, 1986); Peter Klein, *Certainty* (University of Minnesota Press, 1981); Robert Nozick, *Philosophical Explanations* (Cambridge, Mass.: Harvard University Press, 1981), pp. 167—288. Examples of attempts to relate definitions to other topics include: Gilbert Harman, *Thought* (Princeton, N.J.: Princeton University Press, 1973); Paul Benacerraf, 'Mathematical Truth,' *The Journal of Philosophy* **70** (1973): 661—76; Frederick Suppe, 'Afterword,' in Suppe (ed.) *The Structure of Scientific Theories* (2nd ed.) (Urbana, Illinois: University of Illinois Press, 1977), pp. 716—27.

[4] I've considered the basic reliabilist argument against skepticism; there are several variations on it. Some writers, such as Fred Dretske and Robert Nozick, try to capture a *tension* in our way of thinking about skepticism rather than reject skepticism outright. They want to preserve the thesis that we know ordinary common-sense propositions, but also want to cede to the skeptic that we do not know that the skeptic's bizarre hypotheses (e.g. demons, brains in vats, etc.) are false. To achieve this split in intuitions, they work out conditions of knowledge on which knowledge is not closed under known logical entailment, and on which if the world is as we think it is, we know common-sense propositions like "I am working at a desk," but do not know that brain-in-the vat and other skeptical hypotheses are false. For present purposes only the anti-skeptical component of this position will be germane.

[5] Laurence BonJour has made this objection popular. See, e.g. *The Structure of Empirical Knowledge* (Cambridge, Mass.: Harvard University Press, 1985), ch. 3.

[6] Goldman, *Epistemology and Cognition*, chs. 4 and 5.

[7] For further discussion of the possibility argument see the following contributions to Marjorie Clay and Keith Lehrer (eds.), *Knowledge & Skepticism* (Westview Press, 1989): Clay and Lehrer, 'Introduction'; William Alston, 'A "Doxastic Practice" Approach to Epistemology,' pp. 1—30; Barry Stroud, 'Understanding Human Knowledge in General,' pp. 31—50; Ernest Sosa, 'The Skeptic's Appeal,' pp. 51—68. My second parody is related to one constructed by Stroud on pp. 42—44.

[8] Nozick (not without some deviation) presents his view this way, though his argument also includes the points cited in note 4 above.

[9] Sober, 'Psychologism,' *Journal for the Theory of Social Behavior* **8** (1978): 165—91.

[10] Quinton, *The Nature of Things* (New York: Routledge and Kegan Paul, 1973), p. 116.

[11] See Michael Friedman, 'Truth and Confirmation," *The Journal of Philosophy* **76** (1979): 361—82.

[12] Cf. BonJour, pp. 117—23.

[13] Cf. Alan H. Goldman, 'Epistemology and the Psychology of Perception,' *American Philosophical Quarterly* **18** (1981): 43—51.

[14] Sober's argument turns on the greater simplicity of the psychological realist's explanation of the data.

[15] Kim, 'What Is "Naturalized Epistemology?",' in J. Tomberlin (ed.), *Philosophical Perspectives 2: Epistemology* (Atascadero, California: Ridgeview Publishing Co.), pp. 381—405, p. 382.

[16] Swain, *Reasons and Knowledge* (Ithaca, NY: Cornell University Press, 1981), p. 18.

[17] Goldman, p. 36.

[18] I provide a much fuller exploration of these problems in 'Skepticism and Naturalized Epistemology,' forthcoming in a volume edited by Steven Wagner and Richard Warner.

STEVEN LUPER-FOY

RATIONAL DEFINITIONS AND DEFINING RATIONALITY

Beliefs and the processes and procedures used to assess them vary quite a bit from person to person. Analytic philosophers of a certain ilk think they can substantially reduce these kinds of cognitive diversity. They believe that through the process of defining key terms of moral and epistemic justification they can identify and defend principles or procedures for belief adoption, retention, and rejection. These principles and procedures can then be used to identify acceptable cognitive states, thus giving us a rational basis on which to reduce cognitive diversity. Cognitive states are ones that store information about the world; for our purposes they may be thought of simply as beliefs.

However, the analytic project has come under more and more frequent criticism. One of the most important of these attacks has been offered by Stephen Stich, who doubts that through working out analyses or definitions we can generate principles or procedures that will solve the problem of cognitive diversity. His main strategy is to criticize one technique of analysis, namely, the procedure John Rawls calls the "reflective equilibrium" test. Stich's most serious objection to the reflective equilibrium test is that it itself relies on practices and norms that are simply taken for granted by people. But Stich carries the battle to people who do not necessarily use the equilibrium test. Since almost all analytic epistemologists also rely on gratuitous practices and norms in defining critical terms, Stich thinks his "most serious objection generalizes into an argument against an entire epistemological tradition — the tradition that [he calls] 'analytic epistemology'."[1]

I think that the reflective equilibrium test really can help us with definitions of evaluative terms that facilitate good reasoning, the skeptics' doubts notwithstanding. Accordingly, I shall sketch how the test is used, and defend it, taking Stich's criticisms as representative of the project's detractors.

REFLECTIVE EQUILIBRIUM

Suppose that I am trying to decide whether a particular policy is just,

J.H. Fetzer, D. Shatz and G. Schlesinger (eds.), *Definitions and Definability: Philosophical Perspectives*, 203—219.

and that I am not interested in merely stipulating that it is just, nor in stipulating a definition of 'justice' which entails that my policy is just. There are at least three ways to put what I need before I can assess my policy: I need to get clear about what justice entails; I need to analyze the *notion* of justice; I need to define the *term* 'justice.' For my purposes it makes no difference which of these ways of putting matters we choose. I shall speak interchangeably of 'defining the term "justice",' 'analyzing the notion of justice,' and 'spelling out what justice entails.'

To work out what the notion of justice involves, I shall need to provide a set of principles of justice, ones that, at least in the ideal case, lay down necessary and sufficient conditions for a society's institutions to be just. To select these principles, I shall also need an account of the *role* the principles are supposed to play. Once I isolate the role these principles are to play, I can devise a set of principles that will play that role. Following Rawls, we may distinguish between the role principles of justice are to play on the one hand, and the principles themselves on the other, as the *concept* of justice and different *conceptions* of justice, respectively.[2] Thus the principle of utility and Rawls's own two principles are both conceptions of justice, while the concept of justice is whatever these conceptions are *for*. According to Rawls, their purpose is to assign basic rights and duties and to distribute the benefits and burdens of basic forms of social cooperation.

In constructing a complete definition of 'justice', including the provision of both the concept and a conception of justice, I shall have to take quite a bit for granted. As said before, I am not out to simply invent a notion of justice that endorses policies I favor; instead, I presuppose that both a conception and the concept of justice, perhaps more than one, already exist in at least a primitive form, and I want the notion I devise to connect up to those pre-theoretical notions. That is, I presuppose at least a rudimentary idea of what principles of justice are supposed to do for us, and at least sketchy ideas about what sorts of situations are just.

Rawls puts the point this way: he urges me to rely on pretheoretical claims about justice of which I and those in my culture are confident. He calls them 'considered judgments'. These considered judgments which I need to take for granted might spell out considered convictions about what principles of justice are for, as is done, for example, by the claim, 'Principles of justice are to assign basic rights and duties and to distribute the benefits and burdens of basic forms of social cooperation.' In addition, considered judgments might speak directly to the sorts of situations that are just; that is, they might be claims of the form 'in such and such circumstances (which might be purely hypothetical), so and so would be just.' Taking these judgments for granted, I can defend my definition of justice by showing that it explains my judgments; that is, by indicating that if my constructed principles of justice and my constructed picture of the role they are to play were correct then those considered judgments would be true.

Of the two sorts of judgment, convictions as to the *role* principles of justice are to play are the more crucial and the least revisable. Presumably even the most primitive principles about the sorts of situations that are just were introduced in conjunction with a role they were to play, and derived what justification they had from playing that role. Moreover, once we have an account of the role our conception of justice is to play, we will quickly become more interested in devising principles of justice that do a more and more effective job of playing that role than in seeing to it that the concept and conception we construct explains our considered judgments. Indeed, we will come to have rather different considered judgments as a result of fine-tuning the conception of justice so that it serves the role specified by our concept of justice.

Considered judgments are not all I need to complete my definition. I shall also need to make assumptions that enable me to link conceptions to concepts. In defending a set of principles on the grounds that they play their assigned role, I have to make enough assumptions about the world to substantiate my opinion that following those principles really will achieve my goal in providing a conception of justice.

I can defend a definition of justice if I take my considered judgments for granted, and if I assume enough about the world to substantiate my thoughts about which principles would (probably) do what I want them to do. What, now, about those judgments: can I defend them? No doubt I could defend some judgments by appealing to others, but to defend the lot ultimately I would need to take some account of justice for granted. I would need to say something of the form, 'given that the role of justice is such and such so and so are acceptable principles of justice, and given these principles, my considered judgments are in order.' Thus defending my judgments shares a step with defending my definition of justice: in both I show that my definition explains my judgments.

So in settling on what to believe about the justness of particular situations I must appeal to principles whose defense requires that I take for granted some judgments about the justness of particular situations. In working out a complete picture concerning justice, including a concept and a conception of justice as well as judgments concerning particular situations, I shall have to work from judgments to concepts and conceptions and from the latter to the former. I will be finished when all of these elements are in equilibrium. Hence Rawls's term 'reflective equilibrium'.

The mode of justification invoked by Rawls — the practice of working from an account of justice toward judgments about particulars toward an account of justice — is obviously a coherentist one. Coherence theorists maintain that knowledge does not rest on a foundation of any sort; instead, the justification of a view consists in its being part of a conceptual scheme whose various elements cohere together. The position Rawls seems to favor in *A Theory of Justice* is that our account of justice and our various judgments concerning the justness of various cases are prima facie justified when each is internally consistent and each coheres with the other. For convenience, let us use the term *conceptual coherentism* to refer to a generalized version of Rawls's view, namely, the position that internally consistent analyses and judgments are prima facie justified when they cohere each with the other.

Conceptual coherentism is plausible not only when we examine the context of morality but also when we shift to the epistemic context. To illustrate the point, let us sketch how a conceptual coherentist would analyze epistemic justification.

Such an analysis would require us to provide quite a few materials. First, we would need a concept of epistemic justification, that is, a picture of the role such justification is to play. There is considerable disagreement about what this role is. Laurence BonJour and Alvin Goldman describe it as true belief, while Richard Foley and others say that it is the maximization of the number of one's true beliefs together with the minimization of the number of one's false views.[3] My own view is that maximizing the number of our true beliefs would be too tolerant of false views, and that merely supplementing this aim with that of minimizing the number of one's false views results in a policy that is insensitive to the facts that some beliefs are more important than others which share the same degree of verisimilitude and that beliefs that are not strictly speaking true can be more important than ones that are. From the epistemic point of view, extremely general, accurate, and perspicuous beliefs are worth many times their number in ones that are more specific or incomprehensible, even if the latter are more accurate and the former are strictly speaking false. Thus while beliefs like Newton's first law ("Every body perseveres in its state of rest or of uniform motion in a straight line unless it is compelled to change that state by forces impressed thereon") strictly speaking may not be true, it is more valuable from the epistemic point of view than large numbers of beliefs like 'The book near my left elbow has at least 100 pages' even if the latter *are* strictly speaking true. So even if we could make clear sense of counting beliefs, which is a challenge, it would be misleading to say that the trade-off between truths and falsehoods should be numerical. A better expression of the goal that epistemic justification is supposed to help us achieve is *a complete and accurate understanding of what is the case.* The more complete our views, and the greater their verisimilitude and perspicuity, the better.[4]

More would be required for us to complete our definition of epistemic 'justification'. A second requisite is a *conception* of epistemic justification: a set of principles that lead toward a complete and accurate understanding of reality. Third, we need to make enough assumptions about the world to back up our view that our epistemic principles really will help us achieve the epistemic goal. And fourth, we shall need to assume some judgments of the form 'in such and such circumstances an individual would be justified in believing so and so.' These judgments will help us defend the epistemic concept and conception we adopt. Of course, we can defend these judgments (and the assumptions we made about the world) only on the basis of some account of epistemic justification that we adopt. Once again, working out a complete picture involves working from both ends and reaching a reflective equilibrium.

STICH'S CENTRAL CRITIQUE

Conceptual coherentism leads us to base our definitions of epistemic and moral 'justification' on our own judgments and on those of other people in our culture. We take these judgments for granted in working out those accounts. Stich is unhappy with conceptual coherentism, but he also wants to criticize a related view that he calls *analytic epistemology*, defined as "any epistemological project that takes the choice between competing justification rules . . . to turn on conceptual or linguistic analysis."[5] By "conceptual or linguistic analysis" Stich means a conservative explication of pre-existing concepts.[6] Conservative explications may shade into stipulations of new concepts to replace old ones, but they may not depart radically. Stich means to focus only on explications that do not stray far from the original notions.

Stich's criticism of analytic epistemologists and conceptual coherentists is not that they think they can identify the ordinary conceptions and concepts lying behind the judgments of an individual and that individual's community. They can, Stich would allow. Instead, he wants to object to our relying on those ordinary notions after having done little more than identifying them as our own. Granted, we can identify our moral and epistemic concepts and conceptions, as well as those buried in our culture, but why should individuals take those definitions seriously? For example, suppose we discovered that we shared with others in our community a conception of justice that amounted to the principle that there should be a substantial equality of goods. Surely that is no basis for concluding that justice really *does* require equity. That a particular standard *is* ours (and our community's) does not entail that it *ought* to be ours (or our community's).

Nonetheless, it is possible to make the association between definitions we do have and ones we ought to have look very intimate indeed, and Stich's goal is to undermine one or two of the strategies developed in the epistemic sphere for the claim that a certain sort of 'is' does imply the relevant 'ought.' He spends the bulk of his essay attacking a view associated with Nelson Goodman according to which the justification of my principles of reasoning just is their being in equilibrium with my epistemic judgments. In the following passage Goodman seems to equate the two:

> Rules and particular inferences alike are justified by being brought into agreement with each other. A rule is amended if it yields an inference we are unwilling to accept; an inference is rejected if it violates a rule we are unwilling to amend. The process of justification is the delicate one of making mutual adjustments between rules and accepted inferences; and in the agreement achieved lies the only justification needed for either.[7]

If the Goodmanian view of epistemic justification were correct, then seeking the equilibrium proposed by the conceptual coherentist is precisely the way we ought to go about selecting our notions of epistemic appraisal.

Stich's strategy for assailing the Goodmanian view is to point out daffy principles of reasoning that might pass the equilibrium test (thus showing, in effect, that that test is not in equilibrium with our epistemic judgments). If my beliefs and judgments about reasoning are silly enough, it would be no surprise to find that I am able to establish reflective equilibrium between them and daffy principles of reasoning. Thus some people might manage to get the gambler's fallacy past the equilibrium test. Yet "surely we are not at all inclined to say that a person is justified in using any inferential principle — no matter how bizarre it may be — simply because it accords with his reflective inferential practice."[8]

Stich's claim that the Goodmanian view might underwrite daffy rules is only the first of three criticisms he levels against the policy of deciding which notions of epistemic evaluation we ought to have by simply retaining the ones we already have after clarifying them or after adjusting them through the equilibrium test. The second charge is that this policy makes our evaluative notions too arbitrary. Before we conclude that we ought to retain a standard we have identified as our own, shouldn't we take further steps? For example, shouldn't we make sure that our epistemic standards are better than the many diverse and competing standards others might or do adopt? (And shouldn't we worry about competing standards in the moral realm as well, Stich might have asked?)

> The analytic epistemologist's effort is designed to determine whether our cognitive states and processes accord with our commonsense notion of justification. . . . Yet surely the evaluative epistemic concepts embedded in everyday thought [may] vary from culture to culture. Moreover, the analytic epistemologist offers us no reason whatever to think that the notions of evaluation prevailing in our own language and culture are any better than the alternative evaluative notions that might or do prevail in other cultures. [So] why should we care one whit whether the cognitive processes we use are sanctioned by those evaluative concepts?[9]

Stich's third complaint is that only xenophobes would take their evaluative concepts seriously without first demonstrating the superiority of those notions. Analytic epistemologists think that

> When we know that a certain cognitive process falls within the extension of our ordinary terms of epistemic evaluation . . . we know all that . . . is relevant to the questions of how we should go about the business of reasoning. But as I see it, the only people who should take this information to be at all relevant to the question are the profoundly conservative people who find intrinsic value in having their cognitive processes sanctioned by culturally inherited standards. . . .[10]

Their xenophobia leads analytic epistemologists to believe that the epistemic notions of their culture cannot be improved upon. This is a disastrous mistake, Stich thinks, for

> There are *lots* of values that are both widely shared and directly relevant to our cognitive lives, though they are quite distinct from the "epistemic values" that lie behind our ordinary use of terms like 'justified' and 'rational'. It is against the background of these non-epistemic values that our socially shared system of epistemic evaluation can itself be evaluated.[11]

Since, for example, many people value happiness, power, or other things, "we can ask whether the cognitive states and processes we actually have or use foster happiness, power, or the rest."[12]

XENOPHOBIA

In saying that it is xenophobic to take for granted the concepts and conceptions of our linguistic community, Stich assumes that people who take these things for granted want to conform to whatever concepts and conceptions happen to be their culture's. These conformists may consider their loyalty intrinsically valuable or instrumentally valuable. Either way, people who desire to honor nothing but the definitions of their community really deserve to be described as xenophobically conservative. Were a shift in that community's definitions to occur, its xenophobically conservative members would go uncritically along with the crowd.[13]

But Stich is mistaken when he assumes that xenophobia must be responsible any time people take their culture's definitions for granted. There are other attitudes that might lead people to take definitions embedded in everyday thought seriously enough to rely on them. Perhaps people simply trust their own notions, and it is the desire to clarify those trusted notions that leads such people to take an interest in ones embedded in their culture: having acquired a good deal of their conceptual apparatus from people who share the same culture, it makes sense for them to think they can clarify many of their own notions by clarifying those of their peers.

The assumption that our own definitions are unproblematic enough to rely on, coupled with the assumption that many of our notions are roughly the same as those of our peers, will lead us to take an attitude that might well be called 'conservative'. But this is an accidentally conservative attitude, and one that is not xenophobic at all. A change in most of a community's members' definitions would not automatically lead its accidentally conservative members to follow suit. Instead, the latter would drop their assumption that their notions roughly coincide with their peers', and ignore their peers' judgments when analyzing their own notions. If we considered them loyal to something, we would have to say that accidental conservatives were loyal to certain notions that happen to be their community's, not to their community.[14]

So reliance on the definitions of our community without first demonstrating their superiority is not necessarily motivated by xenophobia. If it were, Stich himself could not escape the charge of xenophobia. We have seen that he wants us to evaluate epistemic notions by measuring them against non-epistemic values such as power and happiness. Yet, as he realizes, values like power and happiness themselves might lack any deeper justification. People who are willing to maintain values that are arbitrary in this sense (i.e., not based on anything) need not do so out of the desire to conform to whatever values their community upholds, however. Perhaps they are committed to those values themselves, and would remain so even if everyone else in their community suddenly adopted different values.

RATIONALITY AND STICH'S SUGGESTION

Stich's charge that conceptual coherentists and analytic epistemologists are xenophobic has proven to be a red herring. However, nothing said so far shows that Stich is wrong in his other two charges against the analytic epistemologist's and the conceptual coherentist's conservative practice of taking undefended evaluative definitions seriously enough to use them to assess cognitive states and processes. Wouldn't we do better to adopt Stich's preferred method of picking evaluative notions, which is to select those that are suited to accomplishing fundamental non-epistemic goals such as being happy, powerful, etc.?[15] Before we consider Stich's remaining two charges, let us pause to consider what Stich's idea of rationality must be given that he wishes to assess evaluative definitions by seeing which ones are instrumental to our non-epistemic goals.

The sort of rationality Stich has in mind might be called *generic* rationality. It should be distinguished from what sometimes is called *epistemic* rationality. The former is the efficient pursuit of our goals whatever they are (after we take familiar measures such as giving priority to goals that are more important to us than others, minimizing inconsistencies, etc.). Fully epistemically rational persons, however, must have, or behave as if they had, the epistemic goal, which, we have assumed, is the desire for a complete and accurate understanding of what is the case. In fact, they are people for whom the epistemic goal is always of the greatest importance; their first priority is to do everything they can to achieve it, and they concern themselves with other things only if doing so does not interfere with their single-minded pursuit.

By way of comparison, note first that both sorts of rationality are matters of degree; how rational we are is determined by how efficiently we pursue our epistemic goal in the one case, or our various goals in the other. In neither case must we *attain* the goal if we are to be highly rational, and normally we will be content to 'satisfice' rather than to maximize.[16] The attainment of the relevant goal is not part of an analysis of rationality in the sense of being a necessary and sufficient condition for rational belief; if it were, obviously nothing less than attaining the goal would qualify us for the epithet 'rational.'

Second, highly epistemically rational people are always highly generically rational. The epistemic goal is an extremely demanding one, so people who let nothing get in the way of their epistemic pursuit will forsake all aims that do not contribute to their intellectual endeavor. They may *have* other interests, but must give weight only to ones that improve their intellectual life, directly or indirectly. Thus they may take time to play games simply because doing so refreshes them and enables them to concentrate better.[17]

Third, since people who score high in generic rationality need not possess (or behave as if they possessed) the epistemic goal, they need not be very epistemically rational at all. This last point bears further discussion; dwelling on it will help us see that while generic rationality has its appeal, it has a down side as well.

Suppose that my only fundamental goal were to be happy; I would care about other things only if I thought it helped me to be happy. My circumstances easily might be such that the best way to keep myself happy would be to construct a completely false picture of what was going on around me. To do so, I would need special epistemic principles that would lead me to the right false views and away from the wrong true views. In giving myself over to such wishful thinking, however, and avoiding the awful truth, wouldn't I be a paradigm case of an irrational person?

I *would* be, but we must keep careful track of the sense in which I would be irrational. Wishful thinkers are *epistemically* irrational. Hence the case of wishful thinkers would be a problem for Stich's method of selecting evaluative definitions if his were intended to be the method we ought to use from the point of view of epistemic rationality. However, it would not a problem if, as seems to be the case, generic rationality were what he had in mind. For wishful thinkers might well be generically rational. Certainly there is no escaping the fact that daffy ultimate goals could call for silly reasoning principles, and eudaemonistic wishful thinkers illustrate the point. But if we focus on the fact that generically rational people are efficiently accomplishing what they set out to, perhaps we can grow to appreciate generic rationality.

Before we become too sanguine about generic rationality, however, we should ask if we can use the same technique (namely, focusing on the fact that being generically rational leads us to accomplish our goals efficiently) to get used to a kind of reasoning that is every bit as odd as wishful thinking, namely the fallacious gambler's reasoning Stich adduced against the Goodmanian approach. Stich deplores this type of reasoning, but it is far from clear that Stich can criticize his gambler from the point of view of generic rationality. It is fairly straightforward that people who commit the gambler's fallacy are epistemically irrational in doing so given that this fallacy has no place among principles of reasoning that are suited to producing a complete and accurate understanding of what is the case. But, to repeat, people who are generically rational need not have the epistemic goal. If their fundamental goals are peculiar enough, they need not care very much at all about constructing an accurate view. Conceivably even someone who desperately wants to be defeated as often as possible at craps may be generically rational, and for such a person the gambler's fallacy would come in handy.

Considering this turn of events, Stich's first argument against the conservative way of choosing evaluative notions is unconvincing. Recall why Stich rejected the conservative Goodmanian view that the justification of my principles of reasoning just is their being in equilibrium with my epistemic judgments: Stich thought it obvious that principles which endorse the gambler's fallacy are unjustified, and so they are — if it is epistemic justification of which we speak. As far as generic rationality is concerned, principles that sustain the gambler's fallacy might be legitimate for someone whose pursuits are wacky enough. So in spite of Stich's worries, perhaps the Goodmanian view would be correct so far as generic rationality is concerned.

On the other hand, the very fact that wishful thinking might pass the equilibrium test for ardent eudaemonists precludes our equating the justification of epistemic principles of reasoning with their being in equilibrium. Since passing the equilibrium test need not involve people taking up the epistemic goal (or acting as if they did), it cannot be equated with epistemic rationality. If what is wanted is a specification of the method people who are maximally epistemically rational would use to arrive at their conceptual apparatus, I would recommend something like the following: work with terms whose definitions help you as much as possible in arriving at a complete and accurate understanding of what is the case. This strategy immediately enjoins that we work with categories that allow us to construct views that are broad in scope. In this way we can help ensure that our picture of reality is complete. To promote accuracy, our definitions should be clear and precise in detail. Also, we should work with categories that we can thoroughly grasp so that we can thoroughly understand our views.

ARBITRARINESS

Let us turn to the second of the grounds on which Stich indicts conceptual coherentism and analytic epistemology. Stich's charge is that when evaluative notions are selected the way conservatives want them to be, those notions are arbitrary. Conservatives can say nothing to justify their principles of reasoning over alternatives.

Conceptual coherentists can defend themselves more effectively against this criticism than Stich lets on. One of the things they do in defining epistemic terms is establish an equilibrium between various principles of reasoning and deliberation implicit in their discourse on the one hand and the roles those principles are to serve on the other. It is easy to overlook this step, however, since conceptual coherentists often cite Goodman as their inspiration, and yet Goodman focuses on reaching equilibrium between principles of reasoning and judgments about acceptable reasoning, thus failing to consider the *concept* of justification, the *role* reasoning principles are to play.

Once conceptual coherentists have adopted concepts and conceptions that fit together, they have the means to say something useful about why they prefer their own methods of deliberating and reasoning to various alternatives. 'I use such and such principles of reasoning,' a coherentist might say, 'since as far as I can tell these principles will bring me closer than any others I know about to achieving the goals I have set for such principles.' To the extent that coherentists can convincingly argue that their conceptions achieve the goals for which they were designed, to that extent they can explain why they prefer those conceptions over alternative conceptions, such as ones adopted in other cultures.

Of course, showing that our conceptions do what they were designed to do will not justify us in having the aims whose pursuit is facilitated by those conceptions. Some might be defended on the grounds that they help us accomplish others, but doing so only postpones the inevitable question, 'Why pursue that whole package of aims rather than some other possible package?' Moreover, we may be able to show that *having* and *pursuing* our aims helps us to accomplish our aims, so that our ground project is self-endorsing, but many other packages of aims would also endorse themselves.

The fact that we must have final aims for which we do not have further reasons does indeed show that our evaluative apparatus is ultimately arbitrary in an interesting way. But it is a kind of arbitrariness from which Stich provides us no means of escape. Generic rationality, Stich's preference, is itself arbitrary in an important way. We can argue that our ultimate pursuits are rational from the generic point of view given that *having* an aim helps us *attain* it, but aside from this fairly empty kind of self-endorsement, we cannot expect to justify our aims using the skeletal framework of generic rationality. From the point of view of generic rationality, reason is the servant of our aims, to paraphrase Hume. Still, if we are disappointed in being unable to provide a rationale for our ultimate aims, our grief should not blind us to the fact that we at least may provide ourselves with a rationale for principles of deliberation and reasoning. Stich seems to overlook the possibility of such a coherentist justification when he asks why we should stick to our notions of epistemic evaluation "in the absence of any reason to think that the locally prevailing notions of epistemic evaluation are superior to the alternatives." The reason he overlooks it is probably that when he speaks of "terms of epistemic evaluation" he has in mind only conceptions of epistemic evaluation, not both conceptions and concepts of epistemic evaluation.

A variant of the coherentist's defense against the arbitrariness charge may be available even when we assess from the point of view of epistemic rationality. To the extent that I have principles that help me achieve the epistemic end, to that extent those principles are justified in the epistemic sense. However, we cannot completely eliminate arbitrariness even if we are fully epistemically rational. The epistemic end itself will remain unjustifiable from the point of view of epistemic rationality. From that stance, reason is the servant of the epistemic goal.

Earlier we noted that there is no guarantee that reasoning principles that meet with the approval of the conceptual coherentist will also be epistemically rational; in this regard the methodology of conceptual coherentism would seem more congenial to those who value generic rationality than to those who favor the epistemic sort. For reasoning principles that do not help us achieve our ultimate goals would be as objectionable to the conceptual coherentist as they are to those who favor generic rationality. Here we have a significant dilemma for someone like Stich who seems to favor generic rationality over epistemic rationality *and* who rejects conceptual coherentism.

GENERIC VERSUS EPISTEMIC RATIONALITY

Stich's argument against the conservative approach to choosing evaluative definitions has run into three major roadblocks. First, Stich's preferred method of choosing evaluative notions seems to call for a picture of rationality — namely generic rationality — that gets along quite well with the conservative approach. Second, he has tried to attack the conservative approach using the case of fallacious gambler reasoning, yet that particular example of irrationality might well be endorsed as sensible by the generic type of rationality to which Stich seems committed. Finally, the type of rationality that is least tolerant of conservatism, and hence most useful from the standpoint of Stich's hostility to the conservative approach, is epistemic rationality, which requires us to avoid choosing principles of epistemic evaluation the way Stich wants us to.

To end the discussion here, however, would be unsatisfactory. Some theses that Stich would probably endorse should be identified and defended, at least in brief; namely, the claims that (1) almost everyone is concerned to pursue a spectrum of ends in a highly generically rational way, rather than to pursue the epistemic end alone; (2) we cannot be criticized convincingly for this decision not to let the epistemic project dominate our lives, and hence (3) it is reasonable for us to adopt principles of reasoning and deliberation that sometimes place limits on the attention we give to matters epistemic in favor of attention to other pursuits. Certainly someone might say that pursuing our other aims leaves us less epistemically rational than we could be, but since other things are at least as important to us as intellectual pursuits, we will be unmoved. The second thesis needs defending and the third clarifying; the first is, I hope, plausible enough on its own.

I do not know of an argument against concerning ourselves *solely* with epistemic rationality and in favor of the broader concern with generic rationality that does not ultimately beg the question. All we can do is point out that narrow intellectuals must neglect their other interests, but the advisability of precisely that neglect is what is at issue. By the same token, however, ultimately any argument that we should drop non-intellectual pursuits and focus instead on the epistemic goal will also beg the question. So I shall give an argument for the second thesis that will have dialectical punch only for those of us whose concerns in life are not limited to intellectual matters. Note that if it were a simple matter to learn everything there is to know, then it would prove generically rational to be extremely epistemically rational as well, so that in practice there would be no need to decide whether one form of rationality should take priority over the other. For the most part, as we all know well enough, knowledge is power; to a large extent, the more complete and accurate our grasp of reality is, the easier it is to do what we want to. However, there is so much to be known that completely giving ourselves over to intellectual concerns would force us to neglect our other concerns. Most of us should therefore limit our intellectual concerns by balancing them off against our other concerns, in effect adopting the standpoint of generic rationality to decide where to limit our intellectual interests in favor of other pursuits.

Let me add, however, that even if most of us would find it irrational to be fully epistemically rational, we are well served by the willingness of others to be as epistemically rational as possible. These people are not screwballs, unlike some who are fully generically rational. The product of the intellectual pursuit is *information*, and that is something we can use even if we would not bother to discover it ourselves. In like fashion, not many of us who want to travel in aircraft would bother to learn to build them. Yet it has been to our advantage that others have devoted their lives to that specialization.

In embracing both intellectual and non-intellectual interests, we will need principles of reasoning and deliberation that sometimes steer our attention away from the project of rounding out our grasp of reality. But this, my third thesis, needs clarifying lest it seem more anarchistic than it is.

The sort of limitation I have in mind is modest. We should let our non-intellectual interests help determine what *sorts* of information we shall seek and how many resources we devote to the search. When our important non-epistemic pursuits hinge on acquiring some type of information, we should seek it rather than other sorts even if the useful sort of information is less important from the epistemic point of view. What we ought never do is deliberately deceive ourselves, in spite of the utility self-deception might have for non-epistemic concerns such as our desire to be happy. Only in special cases can false beliefs help keep us happy long-term, and even in those cases we should not stoop to self-deception. For by 'happiness' we mean something like the long-term satisfaction of our goals. In view of our intellectual and non-intellectual ends, the balanced policy is not to construct complicated principles of reasoning and deliberation that lead us to false view which preserve our happiness. Instead, the balanced policy is to seek useful information rather than simply epistemically important information.

Department of Philosophy
Trinity University

NOTES

[1] 'Reflective Equilibrium, Analytic Epistemology and the Problem of Cognitive Diversity,' *Synthese* **74** (1988), p. 391.

[2] *A Theory of Justice* (Cambridge, MA: Harvard University Press, 1971), p. 5.

[3] Laurence BonJour, *The Structure of Empirical Knowledge* (Cambridge, MA: Harvard University Press, 1985); Alvin Goldman, *Epistemology and Cognition* (Cambridge, MA: Harvard University Press, 1986); Richard Foley, *The Theory of Epistemic Rationality* (Cambridge, MA: Harvard University Press, 1987).

[4] I discuss the epistemic goal in more detail in 'Arbitrary Reasons,' in Roth and Ross, eds., *Doubting: Contemporary Essays on Skepticism* (Dordrecht: Kluwer Academic Publishers, 1990), pp. 39—55..

[5] 'Reflective Equilibrium,' p. 405.

[6] 'Reflective Equilibrium,' p. 396.
[7] *Fact, Fiction, and Forecast*, Third Edition (Indianapolis: Hackett Publishing Company, 1979), p. 64.
[8] 'Reflective Equilibrium,' p. 398.
[9] 'Reflective Equilibrium,' p. 408.
[10] 'Reflective Equilibrium,' p. 409.
[11] 'Reflective Equilibrium,' p. 407.
[12] 'Reflective Equilibrium,' p. 408.
[13] What would they do if their community thought it of the first importance that people form their own opinions and never just go along with the crowd? Since this individualism is (by hypothesis) of the *first* importance, they would have to stop being xenophobic.
[14] It is interesting to note that Rawls would qualify as an accidental conservative by this classification (see the way he characterizes justice as fairness as a political theory in 'Justice as Fairness: Political Not Metaphysical,' *Philosophy and Public Affairs* (1985), 223—251, and in 'The Idea of an Overlapping Consensus,' *Oxford Journal of Legal Studies* 7 (1987), 1—24. In light of the following passage (from *Whose Justice? Which Rationality*? (Notre Dame: University of Notre Dame Press, 1988), p. 4), Alasdair MacIntyre might be an example of a xenophobic conservative:

> Those who resorted or resort to academic philosophy hoped or hope to acquire thereby a set of sound arguments by means of which they could assure themselves and others of the rational justification for their views. Those who resort instead to a set of beliefs embodied in the life of a group put their trust in persons rather than in arguments.

However, the view of practical rationality he eventually develops leads me to think that the above passage is a slip of the pen.
[15] Stich seems to regard truth as a non-epistemic goal, but I cannot be sure he does:

> There are lots of values that are both widely shared and directly relevant to our cognitive lives, though they are quite distinct from the "epistemic values" that lie behind our ordinary use of terms like 'justified' and 'rational'. . . . Thus, for example, many people attach high value to cognitive states that foster happiness. . . . And, on a rather different dimension, many people care deeply that their beliefs be true. ('Reflective Equilibrium,' p. 407)

[16] The notion of satisficing is described by Gilbert Harman in *Change in View: Principles of Reasoning* (Cambridge, MA: MIT Press, 1986), and by Herbert Simon in *Sciences of the Artificial* (Cambridge, MA: MIT Press, 1969). Harman claims (p. 68) that while satisficing is acceptable in the practical sphere, "in theoretical reasoning one would not be justified in making an arbitrary choice of what to believe among competing hypotheses at the same level." I maintain that such satisficing in theoretical reasoning is both acceptable and inevitable (see my essay 'Arbitrary Reasons,' in *Doubting*).
[17] Compare John Cooper's discussion of the intellectualist strand in *Reason and Human Good in Aristotle* (Indianapolis: Hackett Publishing Company, 1986).

PART V

SPECIALIZED CONCEPTIONS

ALAN BERGER

IDEALIZED DEFINITIONS IN PHYSICS AND IDEALIZED DISPOSITIONS

Many philosophers attempt to give a naturalistic account, that is, an account that "legitimizes" any notion by defining or offering an extensional equivalence of any notion in the vocabulary of physics or some other "hard science." Various "naturalistic" accounts make use of *idealized* notions, since idealized notions are used in physics. The most sophisticated use made of idealized notions in articulating a naturalistic philosophy is by W. V. Quine. In his famous dispositional account of logic, Quine must appeal to idealized dispositions that do not always manifest themselves.

In this article, I shall focus on the use made of idealized notions by Quine and other leading representatives of naturalism who must appeal to these notions. I then argue that their use of idealized notions is fundamentally different from the use of these notions in the physical sciences. In fact, I shall argue that their use of such notions cannot be used by anyone restricted to naturalistic notions.

I

We begin by considering how Quine must appeal to idealized dispositions in his account of logic. For Quine, it literally makes no sense to attribute to a foreign speaker the acceptance of a sentence that translates into a contradiction in English or the denial of a sentence whose translation into English is a logical truth.[1] It is this last assertion that we shall call Quine's thesis *T*.

Quine's own defense of thesis *T* rests on a behaviorally defined notion of *obvious*. To say that a sentence is obvious is to say simply that a speaker is disposed to assent readily to the sentence upon query; the notion has no further epistemic import. Logical truths, unlike many other truths, are for Quine, all obvious. This is what gives them the appearance of a privileged epistemic status.

With this notion of obvious, Quine can maintain that "objective" translation (i.e., translation that preserves speech dispositions, in particular dispositions readily to assent to and dissent from certain

J.H. Fetzer, D. Shatz and G. Schlesinger (eds.), *Definitions and Definability: Philosophical Perspectives*, 223—238.

sentences) itself suffices to establish thesis *T*. For such translation will count as the foreign speaker's logical truths those and only those of his *obvious* sentences that translate into sentences we call logically true. It will then make no sense to attribute to a foreign speaker the acceptance of a sentence that translates into a contradicition in English or the denial of a sentence whose translation into English is a logical truth.

An immediate problem for this defense of thesis *T* is that not all sentences that we call logically true are obvious. Many logical truths are not assented to and are even rejected. Quine, of course, is well aware of this fact, and in *Philosophy of Logic* he explains that all statements we call logical truths are either obvious or *potentially obvious*. To say that a statement is potentially obvious is to say that the statement can be "reached from obvious truths by a sequence of individually obvious steps."[2]

These sequences of individually obvious steps may be thought of as eliciting a speaker's higher order dispositions to self-correct his other assenting and dissenting behavior toward potentially obvious truths. We may also view these sequences in a way that does not require that we postulate such higher order dispositions. Instead, we may hold that the speaker's language-constituting dispositions *do not always manifest themselves* and that these sequences help elicit these dispositions to assent to all logical truths.

Be that as it may, the situation may be thought of in the following manner. A speaker may be confused and not *initially* assent to all logical truths. Here Quine would not want to say that the speaker does not *really* accept all logical truths. Otherwise, he could not say that all of logic, as opposed to other disciplines, is objectively translatable. All it can mean to accept *S* is that the speaker is disposed to assent to *S* (or exhibit any other "appropriate" verbal behavior) under certain "appropriate" conditions of stimulation. Since all of logic is obvious or potentially obvious, this means that speakers are presently disposed to assent to any logical truth *when placed under the appropriate or right kind of prompting conditions*. Hence once the conditions for prompting are fully met and a person accepts *S*, then it cannot happen – even in principle – that that person would dissent from *S*. Here, by the *appropriate prompting conditions* under which a speaker is disposed to assent to a logical truth, Quine understands the presentation to the speaker of the appropriate sequence of individually obvious steps.

Thus for Quine thesis *T* cannot serve merely as a useful rule of thumb for arriving at correct translation. Rather he must hold that one is required to adopt thesis *T*. For when Quine's linguist translates the foreign sentence *S* into the logical truth *L*, what he is claiming is that foreign speakers will readily assent to *S* under the *proper prompting conditions*. Hence, it makes no sense to say that, though the foreign speaker did not assent to *S* under appropriate conditions, he still (because perhaps of some further confusion) accepts it. Any sentence from which a foreign speaker is "really" disposed to dissent *under the right kind of circumstances* cannot in Quine's sense be objectively translated into a sentence that we call "a logical truth." Quine's notion of a logical truth thus invokes no other reality than the fact that under the appropriate conditions the sentence is always assented to and that all sentences obtained from it by substitution for lexicon are assented to.

In attempting to construe "proper promptings," Quine is at a serious disadvantage from the start. For Quine the presence of "proper prompting conditions" must be regarded as part of the defining criteria for what it is to accept *L* as logically true. On the other hand, the traditional or common-sense linguist can regard the presence of such conditions merely as part of the evidence by which to determine whether a speaker accepts *L* as logically true. He will agree with thesis *T* and then proceed to employ it as one of several rules of thumb for determining what sorts of promptings will get the right verbal response. But he will drop thesis *T even if applied to an obvious logical truth* if a person continues to dissent from a sentence that gets translated into an obvious logical truth. For although, in general, such dissent would count as strong evidence of mistranslation, he will grant that a person might be confused even under these circumstances. But in practice such a case is so unlikely that this possibility may be ignored when doing practical field work. Thus thesis *T*, for the common-sense linguist, would be regarded as only one of many practical maxims that guides him in his study of language.

Compare Quine's view of the notion of a 'confusion' with the common-sense view. Common sense normally holds a speaker to be confused if he or she gives a wrong verdict or the right verdict for the wrong reasons, where a correct verdict is one which accords with certain norms or rules (or that can be seen to be correct on a priori grounds). That is to say, the common-sense notion of a confusion rests on what a speaker *ought* to say, not on what he is "*really*" disposed to say.

But, of course, Quine cannot allow such normative notions in his empiricist account of language and translation. Instead, he must say that when a speaker does not *initially* assent to a logical truth, he is displaying a "*corrigible confusion.*" And by this, Quine must mean simply that although the speaker when questioned does not assent to this statement, under his present stimulations, yet he is nonetheless disposed to assent to it when placed *under the appropriate prompting conditions.* Here again Quine comes up against the problem of explicating just what is to be understood by "appropriate prompting conditions."

The traditional linguist, on the other hand, views the requirement that proper circumstances be present only as a "rule of thumb" useful in acquiring evidence concerning the speaker's meaning. For on his view, it is possible that a person may never assent to a logical truth under any circumstances. A speaker may be *incorrigibly* confused, and yet we may still have independent reasons for translating a certain sentence he never assented to, as a logical truth. For Quine, however, this situation in principle does not make sense. Under the *right sort of circumstances,* a speaker's disposition to assent to a logical truth must be elicited.

II

Let us proceed now to examine in more detail what Quine understands by appropriate prompting conditions. For Quine, the "appropriate prompting conditions" under which a speaker's disposition to assent to a logical truth L is elicited are those conditions in which we present him with a certain kind of sequence of sentences. And what is this certain kind of sequence? The logician's view is that such a sequence is what is called a *proof.* Presumably Quine, too, has this in mind. That is why where he talks of sequences he cites the fact that "the logic of quantification and identity admits of complete *proof procedures*" [my emphasis].[3]

But the problem is what can Quine mean by a proof (i.e., a sequence of the right kind)? For Quine, a proof must mean only a certain kind of seqence of sentences. This sequence of sentences must *merely elicit* present dispositions (to assent). They cannot be sequences that change dispositions. Otherwise, the speaker's language would not have consisted — prior to our prompting him with these sequences — of speech dispositions to assent to all sentences that are potentially obvious and, thus, to all logical truths. It would be possible for a translation of a sentence in a speaker's (prior) language to attribute to him the denial of a statement whose translation into English is a logical truth, and still be correct. This is contrary to Quine's thesis *T*.

Thus, in order to view dispositions to assent to potentially obvious truths (or dispositions to self-correct) as a part of the language we are translating, we must require that the sequences (of individually obvious steps) that we present to the speaker contain *no new information or evidence sufficient to change the speaker's speech dispositions*. Only then will these be sequences, that, in evoking his assent to potentially obvious truths, *merely elicit* the speaker's speech dispositions, rather than change them. The problem now is to explicate, in Quine's terms, what is meant by the notion of a sequence that contains no new information sufficient to change speech dispositions.

Consider the following explication:

(i) A sequence of sentences *X* contains *no new information* for a speaker *Y* if, and only if, for any member *m* of *X*, either *m* is obvious or *m follows deductively* from obvious members that appear earlier in the sequence.

Then suppose we present a speaker with a sequence of sentences such that (1) its last member is a potentially obvious truth and (2) the sequence elicits a speaker's assent to this truth. According to our definition, the sequence contains no new information if the potentially obvious truths that are members of the sequence *follow deductively* from earlier obvious members of this sequence.

The problem with this definition is: how can we explicate in terms acceptable to Quine, the notion of *follow deductively from*? Quine recognizes that for speakers there are implicational relations among sentences in a language; but he is restricted with regard to how he can interpret these relations. Verdictive regularities are not enough,[4] and yet he is committed to rejecting outright a deductive nexus or entailment relation among statements. This latter notion is one of a family of semantic notions that Quine rejects. Moreover, it is doubtful whether there are behaviorally acceptable notions of 'immediately imply' and of 'imply' that come sufficiently close to capturing what we understand by deductive nexus.

Nonetheless, let us now consider a proposed account of 'imply' that is behaviorally acceptable.

We first define the notion of *obvious implication*:

(ii) $S_1, \ldots, S_n$ *obviously implies* S_{n+1} if, and only if, whenever we are disposed jointly to assent to all members of the sequence $S_1, \ldots, S_n$ and are in fact presented with the sequence, we do under query assent to S_{n+1}.[5]

I suspect that this proposal best fits Quine's intentions. For although we may not assent to a potentially obvious truth if it is presented alone, we may assent to it if, given our present stimulations, we are concurrently presented with the relevant sequence of sentences.

We may then define the transitive notion of 'entailment' as follows:

(iii) *P entails Q* if an only if there exists a sequence starting with *P* and ending with *Q* such that each member of the sequence is an *obvious implication* of the preceding members.

Even so, the notion of 'obvious implication' is not without fault. For present purposes, we shall consider only one objection.[6] A speaker may never assent to a potentially obvious truth. The statement might be too complicated for average, or even above average, speakers to analyze as a logical truth. Indeed, the physiological states of members of even an entire linguistic community may be such that the community is incapable of grasping this truth. Under these circumstances, the sequences would not entail (in any sense in which a speaker's assent is required) all the potentially obvious truths.

III

Thus far, we have required that a sequence of individually obvious steps elicit dispositions to assent to potentially obvious truths, thus allowing Quine to attribute these dispositions to a speaker's present language. Still, a speaker even when presented with such a sequence may not assent to a potentially obvious truth.

But now suppose we no longer require that a speaker must assent to potentially obvious truths when presented with the appropriate sequences. Suppose, instead, that Quine regards dispositions to assent to potentially obvious truths, and speech dispositions in general, as idealized dispositions.

Let us construe the notion of an idealized disposition in a fashion akin to certain idealized concepts in physics. Among such concepts are those of an ideal gas, or a frictionless plane as well as idealized boundary conditions. These concepts are idealized in the sense that, *in principle, they can never obtain.* Similarly, our actual assent to a potentially obvious truth may even in the presence of appropriate sequences never take place. Yet we may still be *disposed* to assent to these truths. Even if we *renege* on accepting previous obvious logical truths, perhaps it can be maintained that in this idealized sense, we nonetheless are disposed to assent to these truths. Moreover, just as idealized concepts have an inductive basis for their use in physics, so may these (idealized) speech dispositions have an inductive basis for their use in Quine's behavioral account of a language. These speech dispositions, then, may be viewed as idealized community-wide speech behavior or speech dispositions, grounded in verbal responses that are inductively determinable causal regularities of the speech community.

Different notions of idealizations are used in science in various ways. In one sense, by idealization is meant the introduction of a theoretical term that is really a place-holder for some otherwise-unspecified predicate.[7] Quine undoubtedly sees speech dispositions themselves as place-holders for hypothetical physiological states or mechanisms. Ultimately, then, these dispositons will acquire their legitimacy when they are explained in physiological terms. But the disposition itself has no significant character of its own, for it is to be replaced by physiological terminology.

> A disposition is in my view simply a physical trait, a configuration or mechanism. It can be a disjunctive physical trait, since like effects can come of unlike mechanisms. What makes it a disposition is no significant character of its own, but only the style in which we happen to specify it. . . . Dispositions to behavior, then, are physiological states or traits or mechanisms. In citing them dispositionally we are singling them out by behavioral symptoms, behavioral tests. . . . To cite a behavioral disposition is to posit an unexplained neural mechanism, and such posits should be made in the hope of their submitting some day to a physical explanation.[8]

Quine's rationale for these place-holders described in dispositional terminology is that, although "usually we are in no position to detail them in physiological terms," behind these dispositions lies an unexplained neural mechanism that ultimately will be a part of a physical explanation. His motives, then, are to seek greater generality of scientific explanation, greater unification of large parts of science, and a deeper understanding of phenomena through more adequate conceptions of how things actually work. Indeed, these motives constitute the standard rationale for introducing place-holder entities into our scientific theorizing.

But ultimate justification for dispositions that function as place-holders for underlying structure rests on our identifying them "only by their behaviour manifestations." Yet there is little justification for identification if the behavior is never manifested. For if we do not assent to potentially obvious truths even in the presence of appropriate sequences of individual steps or if we renege on previous obvious logical truths after being presented with such sequences, how are we justified in supposing that we are disposed to assent to such sentences? Indeed, what do we mean by the supposition? It cannot mean that actual manifested verbal behavior is merely the resultant of other dispositions we are known to have. We have already discussed and rejected this possibility above in connection with the notion of 'tendencies.' Suppose even that we were justified in introducing place-holder predicates and thus justified in postulating an underlying structure or mechanism to account for these dispositions. We still would not be justified in characterizing dispositions as dispositions to assent to potentially obvious truths, as opposed to some other dispositional behavior toward these sentences.

Another kind of idealization in science occurs where the laws of science are not in perfect agreement with the experimental evidence for them. Surely, if physicists were to formulate their laws in strict adherence to what is actually observed about physical phenomena, these laws would be statistical in form rather than universal. Galileo, for example, would have to say that the velocity of falling bodies varies with their weight and shape if he were to confine himself to correlating observed data. His laws would have to be statistical rather than universal in form, since "there is only a high correlation rather than an invariable porportionality between the distances bodies fall and the squares of the lapsed times of their fall."[9]

There are several additional reasons for formulating laws in universal form rather than in statistical form, apart from the search for generality, unification and understanding. First, like idealized concepts, universal laws are simplifying devices that lend themselves to a relatively simple formulation, using the available methods of mathematical analysis. Second, experimental evidence often shows that existing things possess traits that are at least not measurably distinguishable from what the universal laws assert about these things. But this second justification surely cannot be used to attribute to a speech community dispositions to assent to potentially obvious truths. For, even in the presence of appropriate sequences of individually obvious steps, speakers may not assent to potentially obvious truths. In that event surely, actual speech behavior is significantly distinguishable from the behavior involved in the positing of these dispositions.

There is still another sense in which the laws of physics may be regarded as idealizations. These laws may be universally valid only under idealized boundary conditions or "pure cases" of the phenomena investigated. Neither the pure cases nor idealized boundary conditions *need ever actually exist, even in principle.* This method is usually employed in order to secure generality of statement, in preference to formulating several laws applicable to only special circumstances.

Galileo's law for freely falling bodies, for example, is formulated for bodies moving in a vacuum, even though bodies always move through some resisting medium. The laws of thermodynamics are formulated with reference to idealized gases. Still other laws are stated for such idealized conditions as perfectly rigid and homogeneous bars, frictionless planes and so on, even though actual objects and situations only approximate these conditions.

In the preceding case, laws were said to be idealized in the sense that they are approximated by experimental evidence. But here, actual observed physical phenomena may deviate considerably from behavior called for by these laws. The distances bodies fall are not invariably proportional to the squares of the lapsed times of their fall, contrary to Galileo's law. This proportionality may vary immensely depending upon the weight and shape of the bodies and other air resistance factors.

Since actual physical phenomena may deviate significantly in behavior from the idealized cases for which the laws are formulated, additional assumptions or postulates are required in applying these laws to actual situations. These assumptions or postulates must account for the discrepancy between the ideal case for which the laws are stated and the actual circumstances to which the laws are applied. But such additional assumptions are often quite complicated and may lack the precision in formulation of the laws. Often we do not know all the factors relevantly distinguishing the actual from the ideal case. Moreover, many of these factors are taken for granted, since explicit mention of all the assumptions would be too onerous. For these reasons, the assumptions may not all be fully stated.

It is under this last sense of idealization in science that dispositions to assent to potentially obvious truths appear to find a place. Actual situations hardly ever realize these idealized dispositions. Moreover, actual community-wide speech behavior may deviate significantly from behavior that accords with the laws or regularities governing idealized speech dispositions. As in the example of Galileo's law, additional assumptions are required to account for the discrepancies between the ideal cases for which the laws or dispositions are stated and the actual circumstances to which the laws or dispositions are applied. Such assumptions may include various factors that may cause a speaker to be in error. At times even dispositions to self-correct, usually evoked by appropriate sequences of individually obvious steps, may not be elicited. As in the case of physics, these additional assumptions may be quite complicated and we may not even know all the factors relevantly distinguishing the actual from the ideal case. Some of these assumptions may resist explicit formulation. For these reasons, it may be argued that postulating idealized dispositions to assent to potentially obvious truths is in principle no different from postulating idealized conditions or "pure cases" in physics.

The idealized condition under which Galileo's law holds is that the freely falling bodies move in a vacuum. The analogous idealized condition under which the laws governing speech dispostions hold is that of a speaker's perfect knowledge of both the grammar (or at least the lexicon) and the set of sentences true in his language. Quine holds that, given these two parts of a language, logical truths can be specified recursively as the true sentences whose truth is preserved under any substitution for lexicon. Accordingly, logical principles can be specified by stating that any sentence of a certain grammatical form is true.

It now appears that Quine can defend his thesis *T*: that logical truth is invariant under a translation manual that preserves all of our speech dispositions that hold under idealized conditions. Any translation from a foreign language would impose our logic on a foreign speaker, if this foreign language is translatable at all. It would do so by counting as his logical truths his obvious (or potentially obvious) sentences that are translated into our sentences that we call logically true.

Such translation would proceed as follows. Since there are an infinite number of sentences in a natural language, we cannot explicitly pair sentences of English with sentences that are their translations in the foreign language. Instead, we need the recursive machinery of the grammars of the two languages. These grammars provide lists of words and grammatical constructions of each language respectively, and a translation associates the words and constructions of one language with those of the other. Sentences of one language will be associated with their translations in the other in virtue of the corresponding pairings of words and constructions, together with grammatical analyses of the sentences into words and constructions. Such a translation would preserve the grammatical construction of our language, and from the grammatical construction, given the class of sentences true in our language, we can specify the logical truths and principles of our language. Thus such a translation would impose our logic upon the foreign language, if such translation is possible at all.[10] It would then appear legitimate to attribute to perfect knowers of the grammar and class of sentences true in their language (and thus attribute to the speech community under this idealized condition) dispositions to assent to all obvious and potentially obvious logical truths.

IV

As I see the matter, there is an essential difference between Quine's idealized conditions and the idealized conditions used in science, and this difference rules out, even in principle, a useful analogy. In physics, we can generally calculate the actual situation from the ideal by means of systematically well-authenticated discrepancies from the ideal. For example, we can measure the coefficient of friction of a surface. With this coefficient number and the laws of motion we can calculate the velocity or the distance traversed before coming to rest of a physical object along this surface. We can independently measure the resistance of media and physical objects traveling in these media. With these measures we can apply Galileo's law of freely falling bodies in order to calculate, say, the distance traversed in an actual situation where these physical objects are falling freely in these media, and so on. These various measures provide systematically well-authenticated discrepancies between the actual and the ideal conditions.

But Quine cannot in principle account, in his own empirical terms, for any discrepancies between the idealized conditions under which speakers assent to potentially obvious truths and actual speech behavior. There is no analogous way to calculate systematically well-authenticated discrepancies between Quine's idealized conditions and the actual ones; we cannot calculate from the ideal case for which the speech regularities of assenting to potentially obvious truths are stated to the actual behavior to which these regularities are to apply.

At most, Quine can supply supplementary conditions that help eliminate the gap between the actual and the idealized behavior where speakers assent to potentially obvious truths when presented with appropriate sequences of individually obvious steps. That is, although Quine cannot calculate the difference between actual speech behavior and the ideal, he might be able to systematically eliminate the "coefficient of friction" in actual behavior. But he can do so only if he is able to exploit the completeness theorem which assures the existence of the appropriate sequences of individually obvious steps. Then he might be able to change speakers' actual behavior into their idealized behavior.

But we have seen that this strategy, in general, fails to change speakers' actual behavior to the ideal. Speakers may not assent to potentially obvious truths even when presented with the appropriate sequences. They may, for example, renege on their previous acceptances rather than assent to the potentially obvious truth. Moreover, should a speaker finally assent to a potentially obvious truth when presented with an appropriate sequence, and perhaps some additional prompting, Quine cannot behaviorally distinguish whether the speaker is eliciting or changing his present speech dispositions. The former, we noted, is required to defend thesis *T*, whereas the latter refutes it.

The problem with this strategy is that Quine does not make use of sequences of individually obvious steps or anything else in order to measure actual verbal behavior from the vantage point of the idealized behavior. Quite the contrary, Quine assigns to these sequences the role of *correcting* a speaker's actual behavior and changing it into the idealized behavior.

In my opinion, there is simply no analogy between Quine's idealized conditions and any idealized notions used in science. These speech dispositions are not idealized in the sense that occasionally a disposition may not be exhibited under any circumstances that normally bring about its manifestation. These dispositions are not idealized in the sense that they are place-holders for something more theoretical. They are not idealized in the sense that actual behavior approximates the idealized behavior. Nor are they idealized even in the sense that these speech dispositions are formulated for idealized conditions or "pure" cases that, in principle, need not ever exist.

The truth of the matter, as I see it, can best be brought out by looking at the role Quine assigns to sequences of individually obvious steps. As we have just learned, these sequences are not designed to measure the actual deviation from the idealized condition and then to predict actual speech behavior. Rather, they are designed to *correct* a speaker's actual behavior and change it into the idealized behavior. That is to say, Quine is not interested in actual speech behavior except to change it into the verbal behavior that speakers *ought* to exhibit. *They ought to exhibit assent to potentially obvious truths or else they would be committing a fallacy.* In other words, these idealized speech dispostions are *idealized toward a norm.*[11] But there is no room for idealizing toward a norm in Quine's brand of empiricism. Norms are used to justify or criticize behavior; they may or may not be conformed with inductively determinable behavior. That is why normative elements in language must be among the notions about language that Quine rejects. He cannot accept norms in language any more than he can accept analyticity or rule-governed behavior or statements in language true by convention — all of which have normative elements. These are notions that do not make sense in Quine's naturalist philosophy.

If we examine the proposed idealized conditions under which speakers are allegedly disposed to assent to potentially obvious truths, we can better see just where Quine is really idealizing toward normative behavior. Recall that these idealized conditions are: (1) that a speaker (community) possess perfect knowledge of both the grammar (or at least the lexicon) and the set of sentences true in his language, and (2) that the speaker's (community's) actual verbal behavior (speech dispositions) accord with his or her idealized competence. The first condition corresponds to a speaker's idealized competence, the second to his idealized performance.

It should now be obvious that condition (2) bears no relation to a speaker's actual verbal dispositions exhibited in community-wide inductive regularities. The condition merely states the behavior required of a speaker in order to avoid accepting a contradiction or rejecting a logical truth. If a speaker's competence is idealized, he has the competence to determine the logical truths of his language. Thus requiring that actual behavior conform to this *idealized* competence assures that speakers behave in accordance with the norms or rules of their language that determine the logical truths of that language. Now what other reason can there be for such a requirement?

We have already seen that Quine cannot disregard condition (2). Otherwise, speakers still may not have the requisite speech dispositions to assent to all logical truths. But suppose Quine were to reject condition (2) and correspondingly revise thesis *T*. Suppose instead of holding that speakers have the dispositions to assent to all (classical) logical truths, he holds instead that speakers have the competence to accept all (classical) logical truths. Here, too, Quine would not mean that they actually have this competence, but, rather, they have this competence under idealized condition (1).

Surely this revised thesis is non-Quinean once we replace talk of speakers' verbal dispositions with talk of speakers' competence. As we have shown above, the use of such a notion would defeat the whole purpose of a dispositional account of language as well as be unavailable to him since he denies it to his opponents. Even so, an examination of this thesis is instructive in that it shows us something about the nature of logic that cannot be accounted for by anyone (even someone rejecting Quine's behaviorism) who holds that logical truths are empirical statements.

The problem with this thesis is that it, too, uses the notion of idealization in a normative sense. The idealized condition relies on making an essential use of language as rule governed: it requires that speakers have perfect knowledge of the grammar of their language. A grammar sufficiently rich in lexicon and structure to specify (by means of the substitution test described above) classical logical truths cannot, as we have seen, be characterized or determined by speakers' speech dispositions or community-wide regularities. Thus the grammar cannot be reduced to mere *laws* or *inductive regularities* concerning the (actual) *use* of the community's language. Rather, the grammar must consist of *rules governing the proper usage* of their language. In other words, the idealized condition requires that speakers have perfect knowledge of the *rules governing how they ought* to use their language, i.e., requires *idealizing toward a norm*. The sentences expressing logical truths in the language *deductively follow* from all the rules or conventions governing the language. It is thus no wonder that knowledge of these rules and how to apply them gives us knowledge of the truths of logic.

Now if one views logical truths as empirical statements, and thus rejects the notion of analyticity, convention or any other a priori constraint governing the use of truths of language, one must reject any idealization of language as rule governed. With such notions one can idealize toward norms and say that speakers of a given language *ought* to accept all logical truths of their language, regardless of whether they do or not. Indeed, the speech dispositions that Quine needs to defend his thesis *T* cannot be characterized as community-wide inductive regularities holding under certain idealized conditions, namely, when these conditions are idealized toward a norm, i.e., toward certain inductive regularities they *ought* to hold. They can never be on a par with idealized laws in physics. In physics we do not say that physical bodies in a vacuum *ought* to fall in accordance with Galileo's law nor do we say that bars ought to be perfectly rigid and homogeneous. In science, the ultimate justification of theories and laws must come from actual phenomena in actual situations, never from norms to which such phenomena ought to conform.

Department of Philosophy
Brandeis University

NOTES

[1] In *Roots of Reference*, Quine modifies his view to apply to only intuitionist logical truths and contradictions. See my 'Quine on "Alternative Logics" and Verdict Tables' *Journal of Philosophy*, LXXVII, # 5 (May 1980) for detail.

[2] *Philosophy of Logic* (Englewood Cliffs N.J.: Prentice Hall, 1970), pp. 82–83.

[3] *Ibid*, p. 83.

[4] See my 'Quine on "Alternative Logics" and Verdict Tables,' *op. cit.*, for detail.

[5] I owe this proposal to Gilbert Harman.

[6] For deeper objections see my article 'A Central Problem for a Speech Dispositional Account of Logic and Language,' in *Logic, Words and Objects: Perspectives on the Philosophy of Willard Quine*, eds. Robert Barrett and Roger Gibson, Basil Blackwell, 1990.

[7] See Ernest Nagel, *The Structure of Science* (Harcourt, Brace and World: 1961), pp. 131–133, for an exposition of this view.

[8] W. V. Quine, 'Mind and Verbal Dispositions,' edited in Guttenplan, *Mind and Language* (Oxford University Press, 1975), pp. 93–95.

[9] Ernest Nagel, *The Structure of Science*, p. 508. See this book, especially pp. 508–509, for a more detailed account of this sense of idealization in science.

[10] See Gilbert Harman, 'Quine's Grammar' in L. Hahn and P. Shilpp (eds.), *The Philosophy of W. V. Quine* (Open Court, 1986), pp. 165–180. (Harman does not argue that Quine should endorse this constraint upon a grammar.)

[11] In his seminar, Kripke criticized the other horn of the dilemma for Quine: if validity is explicated in terms of a speech community's inductively determinable verbal behavior, then we forfeit the ability to distinguish valid from invalid inferences. For, an entire speech community may exhibit and even be disposed to commit many fallacies.

GEORGE N. SCHLESINGER

INVERTED DEFINITIONS AND THEIR USES

I

Consider the attempt to define 'music' briefly as 'an artistic arrangement of sounds in varying melody, harmony, rhythm and timbre designed to give aesthetic pleasure'. I believe that, though it might be improved upon, it is a fair definition, genuinely conveying the gist of the definiendum's meaning and capable of facilitating an understanding of the concept of music. On the other hand, the reverse proposition, consisting in an attempt to define 'sounds' as 'the elements of an artistic arrangement that varies in melody, harmony, rhythm and timbre, so designed as to give aesthetic pleasure' would almost be futile.

From a practical point of view, it could be of help to one who was acquainted with the concept and merely did not know the English term for it. But he who altogether lacked the concept of sound, would *ipso facto* lack the concept of music and might find the definition of minimal use only, e.g., he may infer from it that sounds are not invariably a source of discomfort. But it would not be useful enough even to provide a clue as to how, for instance, distinguish between a floral arrangement and a musical arrangement.

Also, there is little chance that any improved version of such an inverted definition will eventually succeed in conveying what it purports to express, for we cannot capture the idea of music without the concept of sound. Sound is conceptually prior to music; a world may abound with sounds where music does not exist, but there is no music where there is no sound. The notion of music is based on the notion of sound, whereas the latter can stand without the former.

Of the various fundamental rules that govern definitions, one might be formulated as follows:

(D) The definiendum must not be conceptually prior to the definiens.

Violations of rule (D) yield inverted definitions which do not accomplish the usual aim of definitions.

J.H. Fetzer, D. Shatz and G. Schlesinger (eds.), *Definitions and Definability: Philosophical Perspectives*, 239—250.

It is easy to cite any number of blatantly back-to-front definitions. For example, very few would entertain seriously the suggestion that '*X* is a parent of *Y*', should be defined as '*X*'s grandchild is an offspring of *Y*'. There are, however, situations where the inversion is more subtly concealed and is also harder to avoid, as in the context of very basic concepts.

For example, philosophers have gone to great lengths in their efforts to define what turned out to be a fairly elusive concept, namely, 'event' (or 'event identity'), and the issue is still unsettled. Let me cite two of the best-known suggestions that have been made. J. Kim [1] has advanced his well-known proposal that an event be described as the exemplification of a property by a concrete object or set of objects at a given time. On closer inspection this definition appears inverted and thus in violation of (D). Events, such as noises or flashes of light, could conceivably occur in a universe entirely devoid of all concrete objects or substances. There are, however, no possible worlds that contain physical objects and in which no events at all take place. In the complete absence of events, there is no duration, no time. Everything that exists, exists in time. A universe without a temporal dimension cannot contain any physical objects. Thus, events are conceptually prior to physical objects, and the former should not be defined in terms of the latter.

Similar criticism applies to Davidson's famous suggestion [2] about how to define the identity of events:

$$(x = y) \text{ iff } (z)\,(z \text{ caused } x \leftrightarrow z \text{ caused } y) \text{ and}$$
$$(z)\,(x \text{ caused } z \leftrightarrow x \text{ caused } z)$$

where x, y and z are terms denoting events. If this is meant as a basic definition to convey the essence of the concept of an event, then it is immediately seen to be entirely circular: if the concept of an event is not given then we can have no clue as to the nature of causes.

But let us suppose the concept of an event to be given and think of Davidson's definition as purely an explication of the notion 'identical events'. It would still violate (D): completely acausal universes seem possible, that is, there are possible universes where all occurrences are random occurrences, and in such a universe an event x may be identical to y. But it would be absurd to suggest the converse, namely, that a universe in which causal relations are exemplified but no events ever take place could exist. Thus the notion of 'identical events' is conceptually prior to that of causes. In addition to all this, it is hard to see how one gets hold of the idea that the z which caused x is the same z which caused y if one does not already possess the concept of event-identity.

That philosophers should encounter such difficulties in trying to tackle this particular issue is understandable, as it is bound to be a hard task to find anything that is conceptually prior to an event. Let me, however, sketch a possible approach — guided by principle (D) — to the order in which concepts should be defined. It seems appropriate to define physical objects in terms of properties, e.g., as being bundles of properties; properties in terms of events, e.g., an object has a given property if it has a disposition to give rise to such an such an event. Causes too are to be defined in terms of events. With the need to define 'event', we have of course reached pretty near rock bottom. It seems, however, a viable suggestion to try and define it as any change in a physical system during a certain interval of time. A physical system need not presuppose an idea about objects or substances; the term may be taken to refer to any region of space, whatever it does or does not contain. Thus a region of space R which is discernible at times t and t' has undergone change, and that change constitutes an event occurring in R, during $t' - t$. This suggestion does not go as far as to define 'event' in anything that is conceptually prior to it, but neither is it an inverted definition, since a universe devoid of regions of space and changes which contains events seems impossible. Causes are to be defined in terms of events.

II

Principle (D) applies in the context of the standard uses of a definition, which is, roughly speaking, to decribe the past or future meaning attached to a term or to facilitate the grasping of a concept. That is, either as J. S. Mill would have it, a definition is to convey the meaning of a word which it bears in common acceptation or that which speaker or writer intends to annex to it; or as Aristotle said, it is the account of the essence of things. However, a definition which violates (D), or indeed most of the rules governing the standard uses of definitions, may still play all sorts of useful roles (like any meaningful sentences may), roles which are far removed from what definitions are normally called upon to fill.

Let me begin with a brief example. Many contemporary philosophers hold that to understand a statement amounts to being able to recognize the conditions under which that statement is true or false. Quine, for example, in a brief statement has expressed support for the idea of defining 'understanding' in terms of 'truth-conditions,' as when he said: "... a man understands sentence in so far as he knows its truth-conditions" [3]

It is to be noted that Quine's man *knows* the truth-conditions and is not said merely to be able to describe them. One might suggest that it is, after all, logically possible for a man to be able to describe all the truth-conditions of a sentence without actually understanding it or any of the conditions he has uttered. On the other hand, to know the truth-conditions requires justified belief, which does imply understanding.

This however raises the objection that Quine's definition is seriously flawed, for knowledge itself is to be defined in terms of belief and justified belief, and even plain belief — let alone justified belief — presupposes the notion of understanding. One cannot believe what one does not understand at all. Thus the definition is circular. An additional difficulty seems to arise from the fact that there exist indefinitely many empirical statements that one may understand without believing either them or their denial, whereas it is not possible to believe a single empirical sentence without understanding it. Thus understanding may be said to be a more primitive concept and hence prior to belief.

None of this stands in the way of the considerable non-standard usefulness of the proposed definition. Among other things, it provides us with an effective criterion for determining degrees of understanding. As we know, understanding comes in any different number of degrees. R. Clausius, who enunciated one of the most fundamental laws of physics, 'The entropy of closed systems increases', certainly understood that sentence. Yet his understanding cannot be compared to the way it was understood with the advent of statistical mechanics. Quine's definition, which expresses the intimate connection between understanding and truth-conditions, offers a ready way of explaining precisely why. Maxwell and Boltzman understood the Second Law of Thermodynamics better than its originator, Clausius, because the set of truth-conditions of that law known to Clausius was a proper subset of the truth-conditions known to Maxwell and Boltzman.

III

Most of the remainder of this paper deals with what is one of the most impressive illustrations of a non-standard use of definitions I know of. It involves a definition of 'morally obligatory'. But first let me make a few preliminary remarks.

As a practical project it would, of course, be absurd, but in principle we could grade the entire human population according to the amount of moral excellence they possess or lack. Let us imagine that we have a set which contains 10% or 15% of this earth's human inhabitants who are morally superior to the rest of us. Let us assume that we have a full record of the behavior of each member who belongs to this august body of individuals, and note what happened when in the course of the last five or fifteen years any one of them found himself in a situation in which *B* obtained. Suppose we find that a substantial majority of these individuals have, on all, or a great majority of all such occasions, acted so as to ensure the truth of *A* and that in doing so they were at least partially motivated by moral considerations. 'Moral consideration' in this context does not have the limited implication of an action having to derive from some sophisticated, systematic theory of ethics, but a wide spectrum of possible moral motivations, e.g., doing *A* was accompanied by sentiments of compassion; the agent felt shame when contemplating not doing *A*, or he decided to do *A* after considering that it is how an acquaintance revered for his high moral standards would have acted under similar circumstances. I submit, that it is very highly probable — and therefore one is rationally justified in asserting it — that when these conditions are fulfilled, *A* amounts to a moral obligation *O*, given that *B*. I would also maintain the converse, namely, that whenever $O(A/B)$ holds, it will turn out that a substantial majority of the set of morally superior individuals almost invariably acted in accordance with A.

Two comments before continuing. First, is it preferable to speak of a substantial majority rather than a plain majority? Many will agree that there exist acts which, though morally desirable, are not obligatory, and that those who fail to perform them are not to be condemned for moral laxity. Virtuous people are however likely to do more than what is strictly obligatory; they often go beyond the call of duty to perform acts of benevolence. But, of course, for acts that are strictly obligatory, even a greater majority may be relied upon to perform. Since my objective is to set up a system of theorems involving not what is merely morally commendable but what is mandatory, I shall focus on what a substantial majority of right-minded people do.

Secondly, I included the phrase 'they are at least partially motivated by moral considerations'. Without this proviso all sorts of neutral acts would qualify as morally obligatory. For example, a substantial majority of all people, including the majority of all good people, carry an umbrella when it rains, yet doing so satisfies no moral obligation (unless of course it is done for some altruistic purpose, e.g., carrying an umbrella and constantly looking to see if there is anybody requiring protection from the rain). In addition, there are a number of philosophers who would insist on the inclusion of this proviso in the definition of a moral act quite independently of the considerations cited here. For example, P. Nowell-Smith, in a highly influential article, maintained that for an act to qualify as a moral act, the agent has to perform it 'because of the intrinsic desirability of the act', and not because of fear of retribution or harm nor because of expectation of reward or for any other ulterior motive [4].

Now I propose the following definition:

A is morally obligatory, given that *B* obtains, if and only if, the probability that an individual picked at random from the set of righteous individuals who have ever found themselves in circumstances in which *B* applied acted in accordance with *A* (for moral reasons) is greater than or equal to *n*, where *n* equals some fraction between 1/2 and 1.

In symbols:

$$(\delta) \qquad O(A/B) \rightarrow \{P(A/B) \geqslant n\}.$$

Definition (δ) is, of course, flagrantly inverted. The universe could be devoid of a single righteous individual but moral obligations would not thereby lose their force. And, from a practical point of view, (δ) is completely useless: one who has only this definition to go by will not be able to judge of a single act described to him whether its performance is a sacred duty or a heinous crime. And yet it may be found remarkably effective in serving as a criterion by which to determine the status of any putative theorem of deontic logic.

IV

There is no doubt about it: such a criterion should prove invaluable. In the course of its relatively short history, philosophers have advanced half a dozen or more contrary systems of theorems in the logic of moral obligations. Given that at most one of these may be correct, a sufficient number of invalid argument forms are in circulation to provide the right conditions for paradoxes to arise. And this branch of logic is notorious for being plagued by a high number of recalcitrant puzzles and paradoxes. Unlike the various other methods suggested, (δ) provides a universal criterion for determining the status of any putative law of deontic logic. It will not fail to help us distinguish genuine theorems from their plausible imitations. Confusing one with the other is, after all, at the root of all the paradoxes. The following constitutes but a small fraction of the useful application of (δ).

Example I

David Lewis [5] has advanced a system of deontic logic, which he claimed to be both sound and complete, and which is based on thirteen axioms. One of his axioms is:

(a) $O(A/B) \,\&\, O(A/C) \rightarrow O(A/B \vee C)$.

There certainly are indefinitely many cases in which (a) applies. In addition, there seems to be a convincing argument why it is valid regardless of what A, B, and C stand for. The antecedent says that A is obligatory both when B is true and when C is true. If so it should seem obvious that when either B or C or both B and C are true, A is obligatory.

It is, however, possible to construct counterexamples to (a). In the Australian outback, there used to be a Flying Doctor Service which provided emergency medical services for several widely separated, tiny communities. In one remote place the whole service consisted of two physicians, one trained to deal with injuries (to mend broken bones, treat badly burnt people and the like), while the other's expertise was confined to internal diseases. They had a small aeroplane at their disposal which, in addition to the pilot, could carry only one of the doctors and some of his equipment.

We may then state that $O(A/B)$ is true, meaning that given that a message that someone suffered serious injury has been received, the Flying Doctor should take off immediately in response to the call, as well as that $O(A/C)$ is true, i.e., given a message concerning someone who has come down with a dangerous contagious disease, the Flying Doctor should rush to the aid of the patient. Now one day a message arrives saying that in such and such a place an emergency has arisen and the attention of the Flying Doctor is urgently required. The message did not specify — and we shall assume that at the moment it is impossible to find out — what the nature of the trouble was, or the presence of *which* of the two doctors is needed.

Thus, here we are confronted with a situation where we know that $B \vee C$ is true, without knowing whether B is true or C is true. It makes good sense to claim that unless further information is received, the pilot of the Flying Doctors need make no preparations for take-off, since he would be able to fly only one of the two doctors — as likely as not the wrong one, who could do little to help. During his absence another call clearly requiring his sevices would have to go unanswered. Thus, A may not be obligatory given the disjunction $B \vee C$; indeed, it may be even prohibited.

It is not possible to defend (a) by claiming that if we were totally informed we would inevitably know not merely that $B \vee C$ was true, but more specifically whether B or else C was true. Moral obligations, which actually apply to us, cannot be based on purely objective truth regardless of how epistemically inaccessible it may be. Surely $O(A/X)$ may represent a strict obligation; but, as all the available evidence indicate that X is false and the agent is honestly convinced that X is false, he cannot be held blameworthy for not acting in accordance with A, even though, as a matter of fact, X was true. Thus, in general the expression '$O(—/X)$' means 'X is given epistemically to the agent'.

The question what precisely 'epistemic givenness' implies, i.e., with what degree of conviction X has to be held and what degree of warrantedness it must have, would require careful analysis that I shall not go into here. In the present case, the doctors would of course be obliged to obtain as much information as they could. But if, after all their efforts, they learn no more than that some medical help is urgently required, but not specifically which kind, then not knowing which of the two doctors should use the small aeroplane, the pilot need not prepare for take-off, i.e., $\sim O(A/B \vee C)$.

If, however, we make use of (δ), then the realization that Lewis's theorem is invalid will not require us to hit upon a counterexample; it will become evident at once. That Lewis's expression fails to convey a deontic theorem can be seen from the fact that $P(A/B) \geqslant n$ and

$P(A/C) \geqslant n$ do not jointly imply $P(A/B \vee C) \geqslant n$. One of the many ways of demonstrating this is by considering an urn containing six marbles, four of which exemplify all three characteristics A, B, and C; the fifth has B only and the sixth C only. Clearly, since four out of the five B-marbles is also an A-marble, $P(A/B) = 4/5$. Similarly, $P(A/C) = 4/5$. But since all six marbles are either B or C and only four of these are A, it follows that $P(A/B \vee C) = 4/6 = 2/3$. Thus assuming $n = 3/4$, then $P(A/B) = P(A/C) > n$, while $P(A/B \vee C) < n$.

Example II

In an important essay, Hansson [6] lists among the theorems of deontic logic:

(b) $[O(A \vee B/C) \,\&\, {\sim} O(A/C)] \rightarrow O(B/C \,\&\, {\sim} A)$.

Hansson's formula looks very plausible indeed. Given that, when C obtains then the disjunction 'either A or B' is obligatory but A is not, that does not of course imply $O(B/C)$, since the disjunctive obligation may have been discharged through A becoming true. However, given that A is not true, then B is bound to be obligatory. Thus given the antecedent of (b), $O(B/C \,\&\, {\sim} A)$ seems to follow.

When however we examine (b) through our criterion (δ), we find it is not a valid theorem since $P(A \vee B/C) > n \,\&\, {\sim}[P(A/C) > n]$ does not imply $P(B/C \,\&\, {\sim} A) > n$. Suppose there are five marbles in an urn, and all are C, two are A and another two are B. Thus $P(A \vee B/C) = 4/5 > n$, on the assumption that $n = 3/4$; $P(A/C) = 2/5$, i.e., ${\sim}[P(A/C) > n]$, while at the same time $P(B/C \,\&\, {\sim} A) = 2/3 \not> n$.

Indeed, we can imagine situations in which (b) does not apply. In certain countries in wartime, people up to the age of 70 may be called up for service, but those over 55 are not required to do any form of combat duty. Elderly people are given a choice: those, who because of physical frailty or because of a lifelong sedentary job would find menial work too arduous, may choose to perform clerical work; others, who are physically fit and would be very unhappy if they were made to sit behind a desk all day long, may ask to be assigned menial tasks. We may represent all this by $O(A \vee B/C)$, meaning: given that there is a wartime emergency, all persons, in reasonable health, between the ages of 55 and 70 are obliged to do national service, either clerical or menial. At the same time $\sim O(A/C)$ is true (and of course also $\sim O(B/C)$ is true) since nobody is supposed to be forced to do specifically A or specifically B; each individual is to be given the opportunity to choose between the two.

Now suppose that at the moment the circumstances are such that construction of air-shelters has to come to a standstill because of lack of material and no other physical work is required. Would an elderly recruit, who is able and ready to work with his muscles but whom the thought of office work horrifies, be forced nevertheless into clerical work in the absence of any other need? The intent of the law may very well be completely clear on this matter: every elderly recruit must be given a choice between A and B in case A is false, not because the recruit refuses to do A but because circumstances make A impossible; then there is no obligation to do A. In other words $\sim O(B/C \;\&\; \sim A)$ is true, in contradiction to (b).

Example III

This paradox has recently been discussed by Ilkka Niiniluoto [7]. The following set of four sentences seems consistent:

(i) $O(\sim p)$

(ii) $O(\sim q)$

(iii) $O(q/p)$

(iv) p

The first three sentences may arise from the commands:

Don't tell Penrose the secret!

Don't tell Sally the secret!

If you tell Penrose, tell Sally as well!

We shall assume q to be true. Then (iii) and (iv) together imply $O(p)$, which directly contradicts (i).

Niiniluoto thinks that the difficulty is removed through the realization that a system of commands may have a hierarchical structure, so when a higher command is violated it disappears from the system. Thus, once the command not to tell the secret to anyone is violated by making q true, it is cancelled and the conditional command $O(p/q)$ becomes effective.

The idea of a hierarchical structure exemplified by systems of command may be of significance. However, in the present context, it is not needed for there is no paradox to begin with. $O(q/p)$ and p do *not* jointly imply $O(q)$. We realize it as soon as we become mindful of a very important point, which, though fully recognized in the context of straightforward probability statements, is by no means universally acknowledged to apply to moral obligations. Suppose the prior probability of A (= a marble drawn at random, from an urn, of which 1% of the marbles it contains are red, is red) is 1/100. This is compatible with the probability of A, conditioned on B (= the marble drawn was made in Bulgaria) being 99/100. Suppose I draw a marble from the urn without looking at it and am assured that it is a Bulgarian marble, i.e., that B is true. I would then be right to assess the probability of A being red as 99/100. This, however, as we all know, does not cause the value of the *prior* probability, i.e., of $P(A)$, to cease to be 1/100. It only means that, in order to evaluate the posterior probability of A, that is, the probability of A once we are given B, we have to use the equation $P(A/B) = 99/100$.

Thus, assuming it to be morally wrong to divulge a secret one is told in confidence, that is, assuming $O(\sim p)$, we are to assume that the fraction of people among the truly upright who will keep such a secret to themselves is considerably higher than 1/2. Should any individual violate this moral rule, this is not going to change the fact that a substantial majority of hightly decent people do not violate it, which means that $O(\sim p)$ continues to hold. Yet the reprehensible talebearer is now required to ensure that q, because of the rule $O(q/p)$.

The last rule is of course compatible with both $O(\sim p)$ and p; we may infer from it that the great majority of virtuous people who on those rare occasions (because of an irresistible impulse, or a momentary attack of amnesia, etc.) when they commit an indiscretion, i.e., act so as to make p true, may be relied on to follow the instruction to ensure the truth of q.

V

It is to be noted that in both of our illustrations of the non-standard uses of definitions it was not merely permissible, but essential, that the natural order of the concepts involved be reversed. In the first case, we had to treat 'knowledge' as if it was prior to 'understanding', for it is only by first ascertaining the amount of truth-conditions a person knows that we can establish to degree of understanding he has.

In the case of deontic logic, it has proven impossible to reliably determine the status of a putative theorem by inspecting the contents of its numerous concrete instances. Moral situations are so immensely variegated that it often happens that a great number of the examined instances of a deontic rule conform to our moral intuitions; thus, we willingly accept it as a theorem. Yet, as we have seen in the last section, at the same time there also may exist radically different situations – which do not readily present themselves to the investigator's mind – where the reason given for the validity of the theorem fails to apply. To circumvent these difficulties we employ criterion (δ), which enables us to examine a suggested theorem without having to scrutinize all, or indeed any, of its moral ramifications. We proceed by regarding the set of righteous people as given and, through their behavior, we determine the implications of various deontic expressions.

Department of Philosophy
University of North Carolina

REFERENCES

[1] J. Kim, 'Events and Their Description: Some Considerations' in N Rescher (ed.), *Essays in Honor of C. G. Hempel* (Dordrecht, 1969).
[2] D. Davidson, 'The Individuation of Events', op. cit.
[3] W. V. Quine, 'Mind and Verbal Disposition' in S. Guttenplan (ed.), *Mind and Language* (Oxford, 1975), p. 88.
[4] P. Nowell-Smith, 'Morality: Religious and Secular' in B. A. Brody (ed.), *Readings in the Philosophy of Religion* (Englewood Cliffs, 1974), p. 585.
[5] D. Lewis, 'Semantic Analyses for Dyadic Deontis Logic' in S. Stenlund (ed.), *Logical Theory and Semantic Analysis* (Dordrecht, 1974).
[6] B. Hansson, 'An Analysis of Some Deontis Logics' in R. Hilpinen (ed.), *Deontic Logic: Introductory and Systematic Readings* (New York, 1974).
[7] I. Niiniluoto, 'Hypothetical Imperatives and Conditional Obligations', *Synthese*, vol. 66, 1986, pp. 111–33.

PART VI

DISCIPLINARY CONCEPTIONS

MICHAEL D. BAYLES

DEFINITIONS IN LAW

> We are not interested in whether the animal in question is a bird or not in fact, but whether it is one in law.[1]

This quotation is from a Canadian parody of a case — *Regina v. Objibway*. The facts of the case are simple. A poor Indian, Fred Objibway, pawned his saddle and was riding his pony with a feather pillow. The pony broke its leg, and he shot it. He was charged with violating the Small Birds Act which makes killing small birds an offense. The central issue on appeal was whether a pony with a feather pillow on its back is a bird. The act defines a bird as "a two-legged animal covered with feathers." The learned judge's comment above is in response to expert testimony that the animal was a pony. The judge construed the requirement of two legs as a minimum condition. He also held that had the legislature meant to restrict 'bird' to animals naturally covered with feathers, it would have so stated. Thus, for purposes of the act, Objibway killed a bird.

Outside of the contexts of reading and writing, people rarely ask "What does '*W*' mean?" or "What is the definition of '*W*'?" Instead, they ask "What is a *W*?" or "Is this a *W*?" It is always intelligible and often useful to respond, "Why do you want to know?" Unless one knows the purpose for which a definition is to be used, one cannot provide an adequate one. 'Bachelor' means 'an unmarried man'. Does it follow that women and married men cannot receive bachelor's degrees, or that if they do, it is by fraud or theft from single men?

A central thesis of this paper is that the general approach of the judge in *Objibway* is correct; a definition for purposes of law may and sometimes should differ from one for scientific, ethical, or other purposes. Moreover, definitions may appropriately differ for different specific legal purposes. Contexts are crucial. The judge's mistakes are in his rules of interpretation and results — failing to achieve appropriate legal purposes.

A book could and should be written about definition and law. A large part, perhaps too large a part, of this paper involves distinguishing

J.H. Fetzer, D. Shatz and G. Schlesinger (eds.), *Definitions and Definability: Philosophical Perspectives*, 253—267.

various topics and subtopics that might be addressed in such a book. One distinction is between the role of definitions in legal theory and practice. Of course, an account of definitions in legal practice is part of legal theory. Moreover, many terms — contract, property, and so on — play prominent roles in both. The focus here is on definition in legal, especially judicial, practice.

One can ask three different questions about judges' use of definitions. (1) What do they think they are doing? (2) What are they doing? (3) What should they be doing? Although the first two questions are closely related, they are not the same. For example, in *Sherwood v. Walker* Judge Morse appeared to think he was distinguishing the essences of fertile and barren cows, yet he can plausibly be taken to be interpreting contract terms, determining whether there was a meeting of mind, or assigning the risk of a mistake.[2] This paper focuses on the normative question, what judges should be doing.

CONTEXTS AND TERMS

The legal context of definition differs from others. In science, one is usually concerned with understanding — explanation and prediction. Definitions primarily serve to further these aims either directly as part of theory or in developing a theory. For example, in psychology, it can be appropriate to define 'hunger' as 'deprivation of food for *n* hours'. That definition is not appropriate if I am trying to decide whether to have a midnight snack. A purpose of law is to guide people's conduct. Recipes, directions to a party, and ethical norms also serve to guide people's conduct. Distinctively, failure to be guided by law can result in the use of state coercion. In civil and public law, such coercion is rarely used, yet the possibility of its use is always present. The results of failures to follow recipes and directions are "natural" — bad-tasting food, getting lost and missing a party, destroying a computer file, and so on. In ethics, the results are guilt, verbal blame, and so forth.

Suppose Andrews is dying of painful bone cancer. Further suppose that at Andrew's voluntary request, Baker knowingly kills her by poison. Did Baker murder Andrews? In an ethical context, the answer might well be "No," voluntary euthanasia in such circumstances is not wrong. In a legal context, the answer might well be "Yes." We might think that the law cannot permit such exceptions without too great a risk of involuntary homicide. The reverse results might occur if, mistakenly believing Andrews to be alive, Baker shoots Andrews. In law, Baker perhaps cannot be guilty of attempted murder of someone who is already dead, but in ethics there is little doubt that Baker attempted to murder Andrews.[3] Thus, the definitions of 'murder' and probably 'attempted murder' differ in ethics and law.

Within law, the types of definitions one gives and the basis for them should often differ depending on the legal contexts in which they appear. Four general contexts are defining terms from (1) precedents, (2) constitutions, (3) statutes, and (4) legal instruments such as contracts and wills. Precedents, for example, often do not provide authoritative terms to be defined; that is, there is no determinate wording for the holding of a precedent. Even within these different contexts, terms might well receive different definitions. For example, terms in criminal as opposed to civil statutes are to be defined narrowly or strictly.

Finally, it can often be fruitful to distinguish types of terms being defined. One might distinguish between descriptive and legal terms, although the distinction is vague. One might hold a theory of the meaning of legal sentences similar to Hare's prescriptivism or Stevenson's emotivism.[4] If so, then legal terms would have some distinctive prescriptive or emotive meaning. This could significantly affect the type and even possibility of definition. Interestingly, Anglo-Saxon legal philosophers have rarely been concerned with such an analysis of legal terms. Within the descriptive category, one might distinguish between nontheoretical and theoretical terms. Although some philosophers contend that all terms are theory laden, clearly 'ice' is less theory laden than 'volt'. Among legal terms, some are more descriptive than others. For example, 'contract', 'property', 'malice aforethought', and 'murder' are more descriptive than 'right' and 'duty'. The former can be called concrete and the latter abstract legal terms. Concrete legal terms are similar but rarely equivalent to ordinary nonlegal terms. It is a disputable question whether, say, 'intent' or 'duress' should be defined the same within and without law.

REALISM AND ITS ALTERNATIVES

Legal philosophers have taken the usual positions regarding the meaning and thus definitions of both descriptive and legal terms.[5] A realist believes that definitions seek to specify features or kinds existing in reality. A conventionalist tries to encapsulate current (or past) conventions for the use of terms. An instrumentalist holds that definitions should be designed to further certain purposes. A reductionist defines terms by those of another kind. To complicate matters, one might take different approaches to different types of terms. Thus, one could be a realist about nontheoretical descriptive terms, an instrumentalist about theoretical descriptive terms, a conventionalist about concrete legal terms (follow legal conventions), and a reductionist about abstract legal terms.

It might seem that there is nothing special about law to move one to adopt one or another of these positions. That is, the issue has nothing to do with philosophy of law. One's position should be determined by arguments in philosophy of language, epistemology, and metaphysics. If there are natural kinds, then the law should seek to define natural-kind terms in a realistic way. If realism is false, then it is foolish to do so. However, realism obviously does not require (logically, conceptually, practically, morally, or legally) that one provide realist definitions. One can provide stipulative definitions, and sometimes it is a good idea to do so. Thus, even if realism is correct, in law one need not seek to provide realist definitions.

Many legal arguments can be given the following overly simplified structure.

ARGUMENT FORM 1

(1) $d_1, d_2, \ldots d_n$
(2) If $d_1, d_2, \ldots d_n$, then C.
(3) If C, then P.
(4) Therefore, P.

'$d_1, d_2, \ldots d_n$' stand for statements using nontheoretical descriptive predicates. 'C' stands for a statement with a crucial term as predicate. 'P' stands for a legal prescription.

The law should primarily provide conventionalist or instrumentalist definitions of crucial terms.[6] Crucial terms occur in antecedents of legal rules or principles (the difference between them is irrelevant here). They can be descriptive or concrete legal terms. Most definitions in legal practice are of crucial terms. Definitions are needed or wanted to determine whether the fact situation provided by $d_1, d_2, \ldots d_n$ is an instance of the crucial term. The classification of the fact situation matters, because the decision and thus the state's use of coercive power turns on that classification.

An example of this argument form is as follows.[7]

(1) The plaintiff imported tomatoes.
(2) If one imports tomatoes, then one imports vegetables (more simply, tomatoes are vegetables).
(3) If one imports vegetables, then one cannot recover duties paid [statute imposes duties on vegetables but not fruit].
(4) Therefore, the plaintiff cannot recover (or verdict for defendant).

Now suppose realism is correct. Further suppose that the best scientific account of the time, as well as conventional dictionary definitions, define tomatoes as fruit.[8] Must the court follow that definition? Obviously, it need not, because in fact it did not. Should it do so? That depends on the evaluation of the legal norm imposing a tax on tomatoes as well as beets and potatoes but not oranges and apples. there is no a priori reason to believe that it always has the best legal results to classify tomatoes with oranges and apples rather than beets and potatoes even if that does give the best scientific results.

However, Michael Moore has claimed that there is a moral case for using a realist theory.[9] Although he makes similar arguments for specially defined statutory and moral terms, I focus on his argument about natural kind terms of ordinary speech, for if his argument fails for them, *a fortiori* it fails for the others. First, Moore claims that we should use ordinary meanings because they are the best evidence of what the legislature intended, and most people use natural kind terms to indicate natural kinds or events. However, the Court concerned with whether tomatoes are fruits or vegetables concluded, contrary to both dictionaries and scientific evidence, that most people thought of tomatoes as vegetables. The court was plausibly correct in 1883 and perhaps even now.[10]

Second, Moore contends that liberty, fairness, and long-run best results are enhanced by using realist definitions. His examples concern the definition of death. Both cases concern the admissibility of evidence. In the first, it has recently been discovered that people submerged in cold water whose heart and lungs have ceased spontaneous functioning can be revived. Relatives of such a person wish to challenge the determination of death for purposes of removing an organ for transplant. Should evidence of revivability be admitted for construing the definition of death? In the second case, a wealthy woman has become comatose but has spontaneous heart and respiration. The state wants to have her declared dead for the purpose of distributing her estate and thus wants to introduce evidence that irreversible cessation of brain functions prevent her regaining consciousness. Moore contends that once people learn the facts, they would be surprised to have the drowning victim declared dead in the first case. A realist interpretation will enhance the victim's and others' liberty. Moreover, the potential organ recipient will not feel cheated. Finally, in the long run, a realist theory will produce the best results, because it will be based on the best scientific theories.

These arguments about liberty, fairness, and long-run results are not persuasive. First, it is debatable whether it increases liberty and fairness to treat the woman with irreversible cortical brain damage as dead. Suppose she had explicitly requested that she be kept 'alive' in such a situation. If such people are dead, then their wishes are likely to be irrelevant. Second, Moore is concerned with the issue of admissibility of evidence. An instrumentalist would want to admit all evidence affecting the long-run benefits and costs of defining a crucial term one way or another. Hence, on the issue of admissibility of evidence, an instrumentalist will take as broad an approach as a realist. Third, Moore contrasts realism with conventionalism, thus ignoring instrumentalism. An instrumentalist, like a realist, is not locked into a conventional definition.

Finally, Moore takes the *Objibway* parody as evidence for realism. What is wrong with the judge taking a pony with a feather pillow on its back as a bird, Moore claims, is that it conflicts with our knowledge of natural kinds. However, it is as plausible to contend that what is wrong is the legal consequences of so doing. As a result of that construal, an Indian is convicted of an offense for killing a lame pony. If Moore were correct, scientific evidence that tomatoes are fruit should produce the same disturbance. But I, for one, am not at all disturbed by a court treating tomatoes as vegetables for purposes of an import duty. The difference is in the legal consequences, or the evaluation of the resulting rule, not whether the definition coincides with natural kinds. Whether liberty, fairness, and the best long-run results are realized depends on the standards by which judges determine which version of a rule should be adopted, not whether the terms used correspond to natural kinds, assuming there are such.

It might be objected that instrumentalism would result in inconsistent uses of terms in different contexts. Thus, if tomatoes carried fruit flies, one would want to classify tomatoes as fruit for purposes of a statute restricting transportation of contaminated fruit. Realism would provide such consistency. However, such consistency is not found in ordinary language. Terms are used with different meanings and extensions in different contexts. If a builder is asked to save the trees on a lot during construction, the builder is not expected to carefully avoid killing a six-inch pine. Yet, in a nursery, such a pine would count as a tree rather than a shrub. Similarly, a court should classify tomatoes as fruit for purposes of resticting their transportation, but not, perhaps, duties.

The appropriateness of definitions of crucial terms should be determined by their evaluation in legal, not ethical, rules. A striking and controversial example of the difference arises in the Supreme Court's decision on abortion.[11] Speaking for the Court, Justice Blackmun indicates that it need not decide when life begins, that is, whether a fetus is a person for ethical purposes. The Court does find, however, that a fetus is not a person for constitutional purposes. One can argue with the Court's method for determining the legal definition or meaning of 'person', but it still does not follow that it should be defined the same for legal as for ethical rules any more than 'vegetable' should be similarly defined for legal and scientific purposes. Unless one implausibly holds that law and ethics ought to be coincident, one cannot go directly from a definition being appropriate for an ethical rule to it also being appropriate for a legal rule. Consequently, even if realism is correct for ethical terms, it still need not be so for legal ones. One needs a separate argument.

TYPES OF DEFINITIONS

In various places, H. L. A. Hart has criticized the appropriateness of definition by necessary and sufficient conditions of terms in the law.[12] Space does not permit a discussion of his arguments.[13] Whether such definitions are possible or desirable might depend on the type of term involved. However, the use of legal rules does not require definitions by necessary and sufficient conditions. In Argument Form 1, all that is needed is a basis for classifying the fact situation d_1, d_2, . . . d_n as an instance or not of a crucial term. To do so, one does not need a definition providing necessary and sufficient conditions for *C*. The connection needed is "*C if* d_1, d_2, . . . d_n" not "*C if and only if* d_1, *d*, . . . d_n". In *Nix*, the Court did not need necessary and sufficient conditions for something being a vegetable. It only had to classify tomatoes; all it needed was a basis sufficient to classify tomatoes as vegetables. Contrarily, it did not need a sufficient condition for something to be a fruit. It only required knowledge that tomatoes lacked enough properties, or a necessary property, to be fruit.

The search for necessary and sufficient conditions results from a tendency to put legal arguments in a different form as follows:

ARGUMENT FORM 2

(1) If *C*, then *P*.
(2) *C* if and only if $d_1, d_2, \ldots d_n$.
(3) $d_1, d_2, \ldots d_n$
(4) Therefore, *P*.

The logical difference between Form 1 and Form 2 is in the strength of statement (2). Using a modified Occam's razor on premises, one should not use stronger premises than necessary for one's conclusion. Thus, the former argument form is preferable.

Subtle psychological differences can exist between using the two argument forms even though these differences are not logically compelled but perhaps due to the order of premises. In Argument Form 2, especially if premises 1 and 2 are taken as settled, a judge's question is whether a given fact situation comes under the rule. It makes the judge's question sound like an empirical one — does this object have the characteristics of a vegetable? Suppose one is putting groceries in the refrigerator. Form 2 lends itself the following reasoning. "Vegetables go in the vegetable bin. Tomatoes are not vegetables. So tomatoes do not go in the bin." Argument Form 1 tends to make the question whether this fact situation should be handled by this rule, whether the rule should be extended or applied to this situation. The question thus sounds more like a normative issue — should this be treated like a vegetable? Form 1 lends itself to the following reasoning: "Tomatoes easily spoil if not in the refrigerator and are awkward on shelves. Thus, tomatoes should be treated like vegetables and go in the vegetable bin." The second argument, unlike the first, will not work for bananas.

In opposition to definitions by necessary and sufficient conditions, Hart introduced several other types of definitions or terms, including defeasible concepts and excluders. A defeasible term has necessary and normally sufficient conditions for its application, but sometimes even if these conditions are met, its application can be defeated by the absence of other conditions.[14] Hart's example was 'contract'. For a contract, usually there must be two parties, an offer, an acceptance, and consideration. However, even if these conditions are met, a variety of factors can still show that there is no contract, for example, fraud, duress, mistake, and so on. Formally, the form is: C if and only if d_1 & d_2 & . . . d_n, unless e_1 or e_2 or . . . e_n. Central to Hart's claim is the contention that the excuses or defenses e_1 or e_2 or . . . e_n are not simply negative evidence of the existence of some feature such as free and voluntary consent. If the latter were true, then the unless clause could simply become another condition d_{n+1}.

There are objections to the concept of defeasible terms, although some philosophers have happily used it. One could get the logical form of necessary and sufficient conditions by converting 'unless' into 'and not e_1 and not e_2 and not . . . e_n'. To avoid this, one must distinguish positive and negative conditions, contending that the absence of a defense is not a positive condition. Such a distinction is quite suspect.[15] Is absence of an insanity defense in criminal law a negative condition? Insanity is the absence of knowledge of the nature and quality of an act or its wrongfulness, so absence of the defense seems to be a positive condition, namely, knowledge of the nature, quality, and wrongfulness of an act. In any case, all sorts of terms are defined by the absence of some feature, for example, 'bald'. Two species of a genus are often distinguished by one having and the other lacking some property.

For defeasibility, one must thus claim that not all the excluding conditions can be formulated in advance of instances of their application. For crucial terms, especially concrete legal terms like 'contract', this is plausible. The defense of unconscionability has developed in the last forty years or so. This sort of openness, like the possibility of classifying tomatoes as vegetables, most plausibly depends on the normative character or use of the terms. Judges and others came to believe that it is legally unwise to enforce unconscionable contracts.

Excluder terms are basically substitutes for the absence of defenses. Hart claimed that 'mens rea' and 'voluntary' are simply used to indicate the absence of defenses, and Robert Summers took the same approach to the requirement of good faith in contract law.[16] Although I doubt that any of these terms is an excluder, the possibility of such terms cannot be rejected.

Another type of definition can be called a cluster (or paradigm or analogy) definition. With cluster definitions, although some conditions may be necessary, there is another set some but not all of which must also be satified.[17] This might be represented as follows: C if and only if $d_1, d_2, \ldots d_n$ and a subset of $f_1, f_2, \ldots f_n$. A problem with this type of definition is that it might not be worked out exactly what subsets of $f_1, f_2, \ldots f_n$ suffice. Among the conditions $f_1, f_2, \ldots f_n$ some might be more important than others, but still not necessary. With this type of definition, new legal cases can present subsets of $f_1, f_2, \ldots f_n$ not previously decided. Recall that for decisions, courts need not have necessary and sufficient conditions for the application of crucial terms, only sufficient conditions for their applying or not applying. Thus, a prior decision that $d_1, d_2, \ldots d_n$ plus f_2 and f_3 is or is not sufficient for a term applying does not settle situations involving $d_1, d_2, \ldots d_n$ plus f_1 and f_2. Moreover, the previous court need not have specified all the relevant conditions $f_1, f_2, \ldots f_n$. This type of definition is especially useful in arguing from precedents by analogy.

Two other possible types of definitions might be mentioned. One type is definition by example. Legislatures often define terms by listing examples. Perhaps the most common way is to state "including but not limited to" followed by a list. Obviously, the term applies to items on the list, but it is unclear to what other fact situations it might apply. For example, my living will states that if I would not recover from a disability I "not be kept alive by artificial means or 'heroic measures', including, but not limited to, any resuscitation efforts, the transplant of any vital organ, or the use of a respirator." Suppose I were to become irreversibly comatose, would this clause be taken to exclude intravenous feeding? Would one reach the same result were I conscious but physically disabled? Hart has also taken precedents as communicating standards by way of example.[18] However, his claim is perhaps best not thought of as defining terms but as presenting rules — "Do as I do."

Finally, Hart has suggested that legal terms such as 'a right' not be explicitly defined.[19] Adapting Bentham's method of paraphrase, he suggests that one give the truth conditions for a statement using the term 'a right', for example, "He has a right." This fits the tradition that it is sentences, not individual words that have meaning.[20] However, it is probably best to reconstruct Hart's point to the following.[21] 'Clark has a right' does not involve a two term relation as 'Clark has a dog' does. Instead, 'has a right' is a one-place predicate or function. Once 'has a right' is taken to be one expression or function, definition can plausibly proceed by one of the other methods discussed above.

INTENT AND PURPOSE

The preceding argued for an instrumentalist approach to the definition of crucial terms in law. It was also claimed that definitions by necessary and sufficient conditions tend to make judicial decision-making more empirical than normative. If definitions of crucial terms should be instrumental to achieve normative legal purposes, the natural question is "Whose or what purposes should be served?" A complete answer to this question would require at least another article. Here, only a sketch of one can be offered.

The purposes that should generally guide definitions in law are those of a justifiable legal system. This answer appears to open the matter to each judge's opinion as to what would be the best legal system. If so, then a kind of anarchy might prevail. People would not know how terms would be understood. However, this objection contains the seeds of a reply. A legal system that let judges define terms however they thought best would not be the best legal system for the reasons given in the objection. A justifiable legal system will have various constraints and rules for construing terms in different contexts.

Judges should usually construe terms in legal instruments in accord with the intentions of the parties making them. Consider 'artificial means' as it occurs in my living will. Whether that term should be construed to include intravenous feeding were I to become irreversibly comatose should depend on my intent. Determining my intent might pose problems, but they are no different from understanding people in nonlegal contexts (although restrictive rules of evidence might impose obstacles). One can look to what a person has said on other occasions, perhaps even about such matters. In my case, this is easy, because I have written about the value of life when comatose. Limits exist to considering an author's intentions. Words have conventional meanings, often more than one. It is in the choice or moderate extension of one of these meanings that the author's intentions are relevant. One cannot reasonably construe 'artificial means' to include washing my body.

It is not simply because I wrote or authorized the instrument that makes my intentions relevant. Instead, it is because a justifiable legal system would seek to carry out the author's intent in such instruments. In two-party cases, terms are sometimes construed against their authors. If Davis and Everhart disagree about the meaning of a term in a contract that Davis drafted, it might be appropriate to construe it in Everhart's favor. If Davis wanted a term to have a special meaning, he should have made that explicit.

This approach is less plausible for defining terms in statutes and constitutions. Many judicial opinions appeal to the legislature's intent or lack of it. Considering legislative intent here means something analogous to determining an individual's intent in making a legal instrument — references to committee reports, floor debates, and so forth. This is not the place to rehearse the arguments about the existence of anything that might be called legislative intent.[22] There are two central questions. First, is there an interpretation of legislative intent that is sensible and usually permits a reasonable determination of what the intent was? Second, if so, why and to what extent should that intent control the definition? Sensible interpretations of legislative intent are not often likely to provide determinate answers. Rather, they are most likely to help rule out otherwise plausible definitions. The reasons for according them weight pertain to the general power of legislatures to make law. That reason plausibly becomes weaker the more conditions pertinent to the situation have changed.

One can talk of the purpose of a statute or constitutional provision without considering legislative intent as determined by other legislative materials. One way is to determine its purpose from what else the statute or constitution says, for example, by reference to titles, headings, and preambles. Other possibilities exist.

One should be able to develop some principles of statutory interpretation as appropriate for a justifiable legal system. For example, a justifiable legal system would provide persons fair notice of crimes. This is a reason both for sticking closely to ordinary conventional definitions and for construing terms narrowly. Thus, it would be appropriate to define 'vehicle' in a statute criminalizing the interstate transportation of stolen vehicles not to include airplanes.[23]

It might be objected that a surfeit of principles of statutory interpretation exist. Indeed, for practically every principle, one can find an opposite principle.[24] However, that there are many conflicting principles does not imply that a justifiable system would have them or permit them to conflict. First, few people deny that narrow interpretation is appropriate for criminal law. Second, different principles might be appropriate for different contexts. Third, some principles are simply not justifiable, for example, those based on legislative silence. The failure of a legislature to act cannot be taken as a reason for or against something, especially when it never explicitly addressed the issue. Defeating a bill is not a failure to act.

CONCLUSION

Definitions in law raise at least all the problems raised by definitions elsewhere. What is special about law is the normative function or import that definitions have. Normative effects or consequences also exist in eithics, but legal norms are not ethical ones. In view of the normative context of definitions in law, courts should take an instrumentalist approach to defining crucial terms. Even if natural kinds exist and natural-kind terms are to be defined, courts need not and sometimes should not use the natural-kind definition. The judicial function is primarily to decide particular cases. For this purpose, courts need not have necessary and sufficient conditions for the use of a term. All they need are determinations that specific fact situations do or do not justify the application of a term. For this purpose, a variety of types of definition other than necessary and sufficient conditions will suffice, for example, definitions as defeasible, excluder, and cluster terms.

That definitions in law should be instrumental does not preclude preponderant use of scientific or conventional definitions. Indeed, in various contexts it can be important to use such definitions. However, courts usually need to define terms in disputes between two plausible definitions of a term. In those situations, they should opt for the definition best contributing to a justifiable legal system. One should not ask simply which definition would make the legal system best. Instead, one should normally ask what principles of interpretation would a justifiable legal system use for this context, and then use those principles to guide the definition.

Late of
Department of Philosophy
Florida State University

NOTES

[1] Regina v. Objibway, *Criminal Law Quarterly* **8**: 137, reprinted in 'Memorandum,' *Social Theory &Practice* **2** (1972): 197—99.
[2] 66 Mich. 568, 33 N.W. 919 (1887); "Yet the mistake was not of the mere quality of the animal, but went to the very nature of the thing."
[3] There is some doubt whether this situation would be classified as a factual or legal impossibility. If Baker tried to smuggle in a watch mistakenly thinking it was subject to duty, he would not be guilty of an attempt in law, but would be in ethics.
[4] R. M. Hare, *The Language of Morals* (Oxford: Clarendon Press, 1952) and *Freedom and Reason* (Oxford: Clarendon Press, 1963), esp. Ch. 2; Charles L. Stevenson, *Ethics and Language* (New Haven: Yale University Press, 1944).
[5] See Robert Birmingham, 'Hart's *Definition and Theory in Jurisprudence* Again,' *Connecticut Law Review* **16** (1984): 779; Michael S. Moore, 'A Natural Law Theory of Interpretation,' *Southern California Law Review* **58** (1985): 291—301.
[6] Abstract legal terms are rarely defined in legal practice. Thus, the purpose for their definition differs from that given in the text.
[7] Nix v. Hedden, 149 U.S. 304 (1893).
[8] See James L. Weis, 'Jurisprudence by Webster's: The Role of the Dictionary in Legal Thought,' *Mercer Law Review* **39** (1988): 964—65.
[9] Moore, 'A Natural Law Theory,' pp. 322—38.
[10] In grocery stores, tomatoes are usually located with vegetables, not fruits. See also Stephen R. Munzer, 'Realistic Limits on Realist Interpretation,' *Southern California Law Review* **58** (1985): 469—70, who suggests that courts need not interpret 'fish' to exclude whales in a statute limiting the amount of fish that could be caught.
[11] Roe v. Wade, 410 U.S. 113 (1973).
[12] 'The Ascription of Responsibility and Rights,' in *Logic and Language*, ed. Antony Flew, First and Second Series (Garden City, N.Y.: Anchor Books, Doubleday & Co., 1965 [original edns. 1951, 1953]), pp. 152—54; 'Definition and Theory in Jurisprudence,' in *Essays in Jurisprudence and Philosophy* (Oxford: Clarendon Press, 1983), pp. 31—32; *The Concept of Law* (Oxford: Clarendon Press, 1961), pp. 14—16. See also Kent Greenawalt, 'Religion as a Concept in Constitutional Law,' *California Law Review* **72** (1984): 762—67 (definition of 'religion').
[13] For a critique, see Robert N. Moles, *Definition and Rule in Legal Theory* (Oxford: Basil Blackwell, 1987), pp. 55—64.
[14] Hart, 'Ascription,' pp. 154—55.
[15] G. P. Baker, 'Defeasibility and Meaning,' in *Law, Morality, and Society*, ed. P. M. S. Hacker and J. Raz (Oxford: Clarendon Press, 1977), p. 33.
[16] Hart, 'Ascription,' pp. 159—61; Robert S. Summers, '"Good Faith" in General Contract Law and the Sales Provisions of the Uniform Commercial Code,' *Virginia Law Review* **54** (1968): 200—201.
[17] See Hilary Putnam, 'The Analytic and the Synthetic,' *Minnesota Studies in the Philosophy of Science*, ed. Herbert Feigl and Grover Maxwell, vol. 3 (Minneapolis: University of Minnesota Press, 1962), pp. 358—97; see also H. L. A. Hart, 'Theory and Definition in Jurisprudence,' *Proceedings of the Aristotelian Society, Supplementary Volume* **29** (1955): 251—52 (used for 'legal system'); Greenawalt, 'Religion as a Concept,' pp. 767—69 (for 'religion').

[18] *Concept of Law*, pp. 121—22.
[19] 'Definition and Theory,' 34—35.
[20] See H. L. A. Hart, *Essays on Bentham* (Oxford: Clarendon Press, 1982), p. 10.
[21] Birmingham, 'Hart's *Definition*,' pp. 793—94.
[22] See, for example, Moore, 'A Natural Law Theory,' pp. 348—58; Ronald Dworkin, *Law's Empire* (Cambridge: Harvard University Press, 1986), pp. 317—27; and Gerald C. MacCallum, Jr., 'Legislative Intent,' in *Essays in Legal Philosophy*, ed. Robert S. Summers (Berkeley: University of California Press, 1968), pp. 237—73.
[23] McBoyle v. United States, 283 U.S. 25 (1931).
[24] See Karl N. Llewellyn, 'Remarks on the Theory of Appellate Decision and the Rules or Canons About How Statutes Are to Be Construed,' *Vanderbilt Law Review* **3** (1950): 395—406.

THOMAS V. MORRIS

DEFINING THE DIVINE

There is no more central concept in the philosophy of religion than the concept of God. The theist proposes and the atheist denies that this focal concept is instantiated in the actual world. And of course, as in every major philosophical dispute, an understanding and assessment of the best arguments offered by both sides requires a good grasp of precisely what is the subject of argument. Yet it is a rather remarkable feature of the extensive contemporary literature in the philosophy of religion that little attention has been given to the question of a proper definition of the divine. What exactly is meant in traditional religious discourse by those terms standardly translated as 'God'? What is the proper source or method for articulating a philosophically adequate conception of the divine? In this essay, I want to begin to explore such issues as these. As we shall see, the definition of such a metaphysically important term as 'God' is no trivial matter and can itself yield philosophical insight.

(1) OBSTACLES *AB INITIO*

There are of course many different ways to give the meaning of a word. And only some of these can standardly be thought of as consisting in the giving of a definition. Likewise, there are different ways of categorizing definitions. Some definitions are lexically reportive. Others are stipulative. Some definitions are said to be ostensive. Others are clearly and thoroughly conceptual. The word 'God' is obviously a term with a long history of fairly standardized usage. So there is no question of our giving just a stipulative definition of it. Our overall approach must be reportive, and yet we may still find legitimate scope for stipulative refinements, for the sake of philosophical precision.

If we could give an ostensive definition of 'God', much philosophy of religion would be utterly otiose. We cannot give such a definition because God is unfortunately not publicly available to perceptual experience in the requisite way. It has been suggested by some philosophers that the word 'God' can be understood as a purely denotative

J.H. Fetzer, D. Shatz and G. Schlesinger (eds.), *Definitions and Definability: Philosophical Perspectives*, 269–283.

referring expression whose ongoing use is to be explicated by an application of some causal theory of reference.[1] Whether it is possible so to construe some uses of the word 'God' or not, it seems clear to me that the philosophically most interesting uses of the term carry connotative import, and so are amendable to conceptual definition. From there being an object standing in the right causal ancestry to Moses' use of 'YHWH' (in, of course, its original Hebrew form) or to Plantinga's use of 'God', it wouldn't follow that there is a God, and that that particular object is God, regardless of its properties. Most of us would say that if it had none of the properties whose appropriate predicate expressions Moses and Plantinga join together with 'YHWH' or 'God' to form predications, then it isn't God, and so knowing it exists is not knowing that God exists. The issue of whether there is a God, then, is an issue concerning whether certain concepts are exemplified. Is, for example, there a being who is omniscient, omnipotent, and wholly good? Is there a personal creator of this world? The philosophically interesting uses of the word 'God' are then, as I see it, to be thought of as such that what I have referred to as a "conceptual definition" of the word is appropriate, and is to be sought.

But having drawn this conclusion, we immediately face a problem. In the fourth century, the important Christian theologian Gregory Nazianzus once wrote that [2]

> It is difficult to conceive God, but to define him in words is an impossibility.

And in our own day, one religious philosopher has stated that [3]

> the divine nature is eternally beyond the creature's conceptual grasp.

A prominent contemporary theologian adds his conviction that [4]

> God is ultimately profound mystery and utterly escapes our every effort to grasp or comprehend him.

If these worries and negative pronouncements are sound, then it would seem that we face pretty dim prospects for an interesting and adequate conceptual definition or explication of 'God.'

The worry expressed in these passages seems to be that, whatever God is, God must be somehow beyond human conceptualization, from which it would seem to follow that there is little hope of attaining an informative and adequate conceptual definition of 'God.' Fortunately, though, I think that seeing how this worry most often arises will help us to see quite easily that definitional pessimism is not so obviously called for here after all. So let us briefly consider a couple of common roads to such pessimism.

Traditional theists typically characterize God as, among other things, an all-powerful, all-knowing, perfectly good, eternal and transcendent being who has created our entire universe and preserves its existence moment to moment. As creator of the world, it is often added, God is by nature not himself a thing in the world, and so is not the sort of thing we can perceive with our bodily senses. As the bearer of such extraordinary, exalted properties, God would have to be very different from all those things in his created world with which we are familiar. What is of interest here is that each of these quite general features traditionally ascribed to God — his distinctness from any perceptible thing in the world, and his dissimilarity to anything in the world — has given rise to an argument or reason for finally concluding that God must ultimately be recognized to be beyond the conceptual grasp of us his creatures.

Let's first consider briefly the claim that as creator of the world, God himself is not a thing of the world.[5] On one main stream of theological traditions, God is not a thing located in space-time at all. He does not have shape, size, color or hardness. He has no mass or momentum, and is not by nature in any way directly perceptible to the bodily senses through which we acquire knowledge. But those senses are also the means by which we acquire our most basic concepts. Because of this, some theologians have drawn the rash and extreme conclusion that no human concepts can possibly apply to God. They seem to reason roughly as follows, very simply:

(1) Human concepts have been developed to apply to things of the world

(2) God is not a thing of the world

so

(3) Human concepts cannot apply to God.

Of course, the suppressed premise here (generalized),

(sp) If x was not developed to apply to y, then x cannot apply to y.

is transparently false, and the argument is thus accordingly bad.

Other pessimists about defining 'God' seem to think that no human concepts can apply to God because, simply,

(A) God is unlike anything in the world in virtue of his extremely different ontological status, or stature in the hierarchy of reality.

But a single concept can apply to two different things only if those things are alike. Since all human concepts either apply to things in the world, or are built up out of component concepts which apply to things in the world, if God is radically enough unlike anything in the world, and if this is because of what God and things of the world are by nature, then we indeed must conclude that human concepts cannot apply to God, and so a conceptual definition of 'God' looks like a non-starter.

We should first acknowledge that there is an important sense in which (A) is true. God *is* unlike anything in the world, according to all major religious traditions — nothing is his match or equal. But the sense in which (A) is true must be clarified precisely, for (A) can of course be ambiguous between

(A_1) God is not completely like anything in the world.

and

(A_2) God is completely unlike anything in the world.

But (A_1) and (A_2) obviously express very different propositions. Most traditional theists seem to have held that (A_1) is true, but that (A_2) is false. The difference between (A_1) and (A_2) is like that between

(B_1) God does not share *all* properties with anything else.

and

(B_2) God does not share *any* properties with anything else.

(A_1) and (B_1) are compatible with the rosiest propects for defining 'God.' Only (A_2) and (B_2) are such that their truth would prevent our attaining any informative and adequate conceptual definition of God. But of course (A_2) could be true only if (B_2) were true. And it should be obvious that (B_2) is impossible.

If God and worldly things could share *no* properties in common, due to their difference in ontological status, due to the extreme difference in the kinds of beings they are, then, contrary to this supposition, they would then have to share at least one property in common:

(P) The property of having a set of properties not shared by some being with a very different ontological status.

Which is to say, (B_2) cannot possibly be true. Likewise, of course, the claim that no human concepts can apply to God is impossible, in virtue of being self-referentially incoherent. If it were true, then at least one human concept would have to apply to God, namely

(C) The concept of being such as to escape characterization by human concepts,

which is a concept whose very application is incoherent.

But definitional pessimism concerning 'God' need not make such extreme and incoherent claims about the conceptual inaccessibility of the divine. The pessimist could bow to logic and acknowledge the applicability of many basic human concepts to God, such as that of being such that $2 + 2 = 4$, being self-identical, being at least as great as himself, and so on, and yet attempt to insist that the transcendence of a divine being would necessarily place him outside the scope of at least most informative, substantive human concepts.

This is a much more circumspect claim, which does not obviously fall to the problems of incoherence attending the more extreme form of definitional or conceptual pessimism. But it in turn faces a quite difficult epistemological problem. For even this more moderate claim is itself fairly sweeping and weighty, and surely seems to imply or presuppose some important knowledge about God. If it were true that no informative, substantive human concepts applied to God (save any necessary for the expression of this claim itself), how could any of us possibly know, or have good reason to believe, it to be true? It would seem that a certain amount of knowledge about God, or a certain range of justifiable conceptualizations of substance concerning the divine, would be necessary for us to be in any position justifiably to affirm such a stance of pessimium. But of course its truth would seem to prevent us from ever getting ourselves into such a position.

I take it that any severely pessimistic view concerning the adequacy of human concepts for defining the divine will face some such difficulty, logical or epistemological, and that this means that, from the outset, we need not allow ourselves to be dissuaded from the attempt to arrive at such a definition. But whether any measure of success awaits us remains to be seen. So to the positive task we now must turn.

(2) THE PROBLEM OF METHOD

The ancient Greek poet Pindar once posed the question "What is God?", and offered the answer "Everything." Whatever its intellectual merits given the time and place of its utterance, this answer suggested by Pindar was clearly not a very discriminating response. The view which he expressed so succinctly, the view that God is everything or, as it is usually stated, the view that everything is God, is typically called pantheism, which is, at best, a sort of degenerate form of theism. It offers an identification of God, which presumably presupposes a definition of 'God' (or 'Θeos') in accordance with which the identification could be made. Whatever the underlying definition, the identification is, among religious people, a fairly non-standard one. Most religious people use the word 'God,' and its correlates in other languages, not to refer to everything in general, but rather to something in particular, something quite sharply distinct from the limited objects in our ordinary physical environment, something on which we, and they, depend. This is the sort of religious view most properly known as 'theism', the belief in a God distinct from all other things. Any belief that there are two or more such beings, independent of each other ontologically, is a band of polytheism. The belief that there is only one ultimate divine being, the conviction proclaimed in one way or another by Judaism, Christianity, and lslam, is of course the view known as 'monotheism.'

This variety of ways in which the divine has been conceived creates a philosophical problem, and the problem is exacerbated greatly by the fact that even, for example, monotheists have held significantly different ideas of God, ideas that differ in such a way that they cannot all possibly be right. So it seems that from the outset, in our attempt to arrive at a reportive conceptual definition of 'God,' we face a problem. Which, of all the many rival options with which we are presented, is to be selected as the proper idea of God? Is one best? Is one most appropriate or most adequate? It is often convenient to speak of 'the concept of God' as if there is some one and only one idea of God, but there is no single, substantive, detailed concept of deity common to all theistic religions. There is not even unanimity within a single identifiable tradition. Jewish theologians and philosophers disagree among themselves. Christians differ with each other, and not just on trivial points. Some say that God is a temporally everlasting being. Others claim that he is outside time altogether and exists rather in a timeless eternity. Some say that God is altogether immutable, others that he dynamically changes in his interactions with the world. Some ascribe passions to God, others see him as both disembodied and dispassionate, or, to use a more traditional term, they consider him to be impassible as well as incorporeal.

This apparent cacophony of opinions raises a deep question, or set of questions. Where do we get our idea of God, ultimately speaking? Is there some method for articulating an idea of God and, if so, why is there then so much disagreement over the resulting conception? Are there, rather, many different rival methods for arriving at a basic conception of deity? And if there are, which is to be preferred? How is this to be decided? Is there a way of rationally choosing a best method for defining the divine? These are questions about which contemporary philosophers have been strangely silent.

In the introduction to his well-known book *Faith and Knowledge*, John Hick refers to the Old and New Testaments of the Bible as "providing the particular conception of deity to be presupposed" in his discussion, and he explains further that,[6]

> This conception of deity may be briefly, but for our purposes sufficiently, characterized as that of the unique infinite personal Spirit, "holy, righteous, wise and loving," who has created the existing universe and who is fashioning human personalities for eternal fellowship with himself through their own free responses to the environmental challenges and opportunities which he appoints.

And in the introduction to *The Coherence of Theism*, Richard Swinburne begins by stating that, by the word 'God,' a theist understands[7]

> something like a 'person without a body (i.e. a spirit) who is eternal, free, able to do anything, knows everything, is perfectly good, is the proper object of human worship and obedience, the creator and sustainer of the universe.'

This is as explicit as most recent writers have been in indicating what it is they mean to convey by the word 'God,' But, clearly, neither of these passages presents us with a proper conceptual *definition* of 'God.' Neither pretends to completeness or precision. And each acknowledges its limitations. But even apart from this, it would be difficult to take either list of attributes as even the beginning of an attempt at a definition. For one thing, conceptual definitions are standardly taken to consist in, or at least to support for each predicate, something like analytic propositions, propositions whose truth value has the status of necessity and which can be known to be true, in some sense, *a priori*. But there are many attributes ascribed to God in the Christian tradition, attributes which appear in the lists of Hick and Swinburne, which are not thought to characterize God necessarily, and are said to be known not *a priori*, but by divine revelation, or, at least, empirically. It is not, for example, held to be a necessary truth that God is fashioning human personalities for eternal fellowship with himself. Nor is it deemed to be the case that in every possible world God is an object of human worship, since it is not the case that humans exist in every possible world, and it is not the case that God is worshipped by humans in every world in which he exists.

But the Hick and Swinburne accounts are of interest in a number of ways. First, they exemplify the casual approach by which philosophers nowadays introduce the term 'God' into their discussions. Second, they are interesting in that they overlap, presumably, since they pick up on a single tradition. But, third, they are even more worthy of note because they diverge to some extent, and thus remind us of the deeper and more extensive differences to be found among other philosophers writing on the topic of God. Wolterstorff, for example, is one who holds that God is in time. Stump and Kretzmann, on the other hand, see God as outside the space-time continuum. It might seem to be no great challenge to identify a common definition of 'God' underlying the descriptions of Hick and Swinburne. But surely it begins to get a bit more difficult to locate the common conceptual ground among writers who more dramatically diverge in their descriptions of the divine.

It is the recognition of this which presses us to ask what method we shall use to attempt to arrive at a definition of 'God.' I suppose that, in principle, there are many possible such methods. We want a definition which will accord with standard religious usage. We also want one whose elements encompass or support just those analyticities which we intuitively can recognize as such. We want one which will display the core of conceptual common ground between Hick and Swinburne, and also between Wolterstorff, on the one side, and Stump and Kretzmann on the other.

First, we need to identify the range of religious usage relevant to our quest. Divinity amounted to one kind of status for the ancient Romans and Greeks. It seems to amount to something very different in the developed religious traditions of monotheism, especially for the Judeo-Christian stream of religious belief. We shall limit the reportive side of our conceptual search to the more sophisticated traditions. One common typographical way of signaling this is to say that what we are seeking to define is 'God,' not 'god.'[8]

In these traditions, it seems to be widely assumed that divinity can be thought of as something like a status had in virtue of the having of a number of conceptually distinct attributes, and which is such that it is not possible that it be had by a being unless those attributes all are exemplified by that being. From this perspective, predications like

(G_1) God is omniscient

and

(G_2) God is omnipotent

have been commonly thought to express necessities *de dicto*. Nothing, on this perspective, could be God unless it was omniscient. Nothing could possibly have this exalted status unless it was onmipotent. And so on, for various other central predications of what are commonly thought to be the core divine attributes.

We thus need a definition of 'God' which will underwrite such an explication of deity, and which will serve to display the common ground beneath the otherwise widely divergent particular beliefs which can be found within the sorts of traditions that are our selected concern. I believe that we have two main contenders for such a definition to consider, definitions which are associated with two different approaches to theology. They embody distinct touchstones for thinking about deity, but I think they arise out of a common concern and, with its help, can be viewed in a new way.

(3) ARRIVING AT THE BEST DEFINITION

The two different approaches to theology I have in mind can be referred to as 'Creation Theology' and 'Perfect Being Theology.' As I have suggested elsewhere, it seems to me that most contemporary philosophers working in the area of philosophical theology have instinctively taken up the posture and procedures of Perfect Being Theology, even when they haven't reflected or commented explicitly on the fundamental issue of definition at all.[9] This approach is rooted in St. Anselm's famous description of God as "that than which no greater can be conceived." In its contemporary form it understands 'God' to be defined as 'the greatest possible being.' By this, Perfect Being theologians typically mean that God is to be thought of most centrally as a being with the greatest possible array of compossible great-making properties, where a great-making property is understood, roughly, as a property it is intrinsically good to have, good to have regardless of external circumstances. Put another way, to have a great-making property is, so far forth, to have a certain positive status or metaphysical stature which is in no direct way relationally derivative from external circumstances. This core definition of 'God' is then in turn used to provide for a fuller, more detailed conception of the divine in the following way: In considering candidate properties for inclusion in our overall conception of divinity, we first consult our value intuitions to determine, at least *prima facie*, which of those properties are to be considered great-making properties. As our list of such properties grows, we then begin considering questions of property grouping, compossibility, and comparative, additive commensurability. Ideally, the result of such an exercise will be to arrive at an array of divine attributes consonant with the Anselman starting point, the definition of God generative of Perfect Being Theology, an array of attributes which is such that it can reasonably be viewed as both constitutive of deity and possibly exemplifiable.[10] Working in this way, practitioners of Perfect Being Theology have ordinarily attained to a notion of divinity as encompassing personal agency, perfect power, perfect knowledge, and perfect goodness.

It is particularly important to be aware of the heuristic and conceptual dynamics of a process like that constitutive of Perfect Being Theology. First and foremost, we need to note that there is within its procedures a differentiation of levels for thinking about the divine. There is most fundamentally, of course, the core definition of 'God.' Then, by a procedure naturally attached to that definition, a conception of the overall essence of deity, in its conceptually distinct particulars, is developed. The precise details at this level are to some extent open to dispute and negotiable within the practice of Perfect Being Theology. Value intuitions can differ and are in any case defeasible in epistemic status. And whether any particular property or group of properties deemed great-making by our value intuitions is such that it can possibly be exemplified is itself a matter that must be brought before the bar of metaphysical argument and, finally, of modal intuition.

Many critics of theism in the past few decades have alleged that the concept of God is internally flawed and incoherent. They have thought that by merely examining the definition of 'God' we can see that there cannot be any such being and that theism can therefore be rejected on purely logical grounds. This whole strategy, as it has been employed in practice, has evinced misunderstanding concerning the proper epistemic dynamics of any such conceptual elaboration of the divine as we find in Perfect Being Theology. What we have here is an inherently flexible approach. Many arguments have been produced to attempt to show that some understanding of omniscience is incompatible with some conception of omnipotence, or that incorporeality of a certain sort is such as to preclude agency of another sort. Most such arguments just fail on their own terms. But the most that such arguments can typically even hope to secure is the acknowledgement that what we initially had thought perfection to consist in, in detail, it cannot, and thus that our fine-tuned conception of deity in its particulars must be reformulated. What the critics have overlooked is that we can come to think that we have been wrong in exactly what we supposed perfection to consist in without being forced to conclude that there is not or cannot be a God, or that the Perfect Being conception of God is inherently wrong. The problems we may encounter on what we may call the "second level" of concept building here, in our attempt to articulate the specifics of deity, can be for the most part, if not entirely and forever, isolated from the question of the appropriateness of the core definition of 'God' constituting level one of our concept construction.[11]

There have been a number of objections raised against Perfect Being Theology throughout the years, but quite recently it has been argued that all of them can be adequately answered.[12] What has continued to be troubling to some philosophers, however, is the awareness that there exists an apparently rival starting point for theological reflection with equally ancient credentials. This is the perspective of Creation Theology, an approach to thinking about deity remarkable for both its initial simplicity and potential sophistication. A very simple and common form of Creation Theology defines 'God' straightforwardly as 'The creator of this world.' But right away a problem appears. Taking the modal ramifications of a conceptual definition seriously, we seem to be presented here with a definitional truth whose content is highly controversial at best, namely,

(C) Necessarily, God created this world.

Now, some theists in the western tradition have in fact held that, necessarily, a being who is perfectly powerful, perfectly knowledgeable, and perfectly good would create some world or other, in which to manifest his goodness, containing creatures with whom to share his goodness. But only a much smaller minority ever have held, on Leibnizian grounds, that it would be necessary for God to create *this world*, the exact universe in which we live, in all its given particulars. And even if they are right, even if it can be argued persuasively on metaphysical grounds that (C) is true, it would seem altogether inappropriate for us, prior to all such argument, just to produce (C) as an analytic, definitional truth about God, a starting point for discussion.

A slightly more subtle form of Creation Theology, however, can avoid this awkward consequence easily. On this sort of view 'God' can be defined as 'the creator of anything which might exist distinct from himself,' or, in other words, 'the font of all possible existence distinct from himself.' From this we can derive analytically the conviction that, if there is anything which is God, then no universe can exist distinct from him without having been created by him. But this of course in no way entails that in every possible world, or in every possible world in which he exists, God creates the universe in which we in fact now live.

So a sufficiently subtle form of Creation Theology can offer a rival definition of 'God' which escapes any immediately untoward consequences. And it can provide a way of conceiving the divine which accords both with deep religious intuitions as well as with a rich tradition of significant theological reflection. In fact, this perspective on divinity, along with its associated method for conceptual elaboration, seems to appeal as well to those with a somewhat scientific cast of mind. Once this definition of 'God' is in place, it is a fairly straightforward procedure to begin to fill out a more detailed conception of deity. God is being thought of as the supreme causal explanation for the existence of our world, and as the ground of possibility for any other world which might have existed. He is then, along these lines, credited with whatever attributes would need to be postulated in order to secure a good explanation of this sort, given what we know about the universe in which we live, and given our intuitions about other possible worlds. And, again, with this method of conceiving the divine there is a difference of levels of postulation, with respect to epistemic status. Creation theologians can disagree on various particular attributes while yet still sharing the same starting point and overall procedure for conceptual elaboration. Likewise, an attack on any particular postulation, as in the case of Perfect Being Theology, need not translate into an attack on the definitional starting point or on the claim informed by it that there is a God.

If we have here two genuine rivals for a proper definition of 'God', we may be faced with a difficult decision problem. For the strands of thought represented by Perfect Being Theology and Creation Theology seem deeply intertwined throughout the history of Judeo-Christian thought. They also can be thought of as theological echoes of Platonism, in the case of Perfect Being Theology, and Aristotelianism, in the focus of Creation Theology. Both the concern with value and the concern with causation are enduring and legitimate preoccupations within the enterprises of general philosophical reflection and theological inquiry.

It is my suggestion that these apparently distinct starting points for philosophical theology be brought together and their respective methods be conceived of as inextricably bound. In another place, I have argued that anyone who begins with Creation Theology and endorses a few simple and independently plausible metaphysical theses widely endorsed by theists of all kinds throughout the centuries will find himself with the conclusion that God is a, or even *the*, greatest possible being.[13] That is to say, starting with Creation Theology can plausibly be thought to result in the endorsement of the fundamental tenet of Perfect Being Theology, the full explication of which will require the use of its attendant method. I think it is also quite plausible to hold that a reasonable development of Perfect Being Theology will result in an endorsement of the core thesis of Creation Theology, the claim that God is the Creator of anything which might be distinct from himself. For imagine that God is the greatest possible being and that a physical universe distinct from him exists. If we seek to imagine this universe as existing wholly independent of him, as depending on him in no way for its existence or natural activity, it would seem that the resultant conception of God would be of a being who was not after all the greatest being possible. For, surely, a being would have greater value, greater metaphysical stature, if it was an absolute source of existence, such that nothing else could exist without deriving its reality from this being. Thus, in order for God indeed to be thought of as the greatest possible being, he would have to be thought of as what we can call 'an absolute creator.' Such, I think, is the link from Perfect Being Theology to Creation Theology.

And once the Perfect Being theologian has arrived at this point, it stands to reason that he will find the method associated with Creation Theology to be important for filling out that conception of God as absolute creator. If I am right about this, then regardless of one's starting point, it is reasonable to think that one will end up endorsing the core claim and employing the methods distinctive of the apparently rival definitional and conceptual procedure. There is surely something intellectually satisfying about this result. For Perfect Being Theology focuses on the intrinsic properties of God, whereas Creation Theology emphasizes the actual and potential relations holding between God and all else possible. They both seek to explicate one facet or another of metaphysical ultimacy, the intent they both have in common.

It is thus my suggestion that the proper starting point for philosophical theology be thought of as a compound definition. God is, by definition, to be thought of as the greatest possible, absolute creator. And this definition is to be explicated, the concept of God to be expanded, by a dynamic interaction of the two methods of conceptual elaboration associated with Perfect Being Theology and Creation Theology, correctable, of course, within bounds, by any purported data of revelation, plausibly taken to be such. This recipe may give us a more comprehensive, complex, and deeply satisfying way of defining the divine.

Department of Philosophy
University of Notre Dame

NOTES

[1] See, for example, Richard B. Miller, 'The Reference of "God",' *Faith and Philosophy* **3** (1986), pp. 3–15

[2] From 'The Second Theological Oration — On God,' reprinted in *Christology of the Later Fathers*, edited by Edward R. Hardy (Philadelphia: The Westminster Press, 1954), p. 138.

[3] Marilyn M. Adams, 'Redemptive Suffering,' in *Rationality, Religious Belief, and Moral Commitment*, edited by Robert Audi and William J. Wainwright (Ithaca: Cornell University Press, 1986), p. 251.
[4] Gordon Kaufman, *God the Problem* (Cambridge: Harvard University Press, 1972), p. 95.
[5] Much is made of this claim by some recent writers influenced by St. Thomas. See, for instance, Brian Davies, *An Introduction to the Philosophy of Religion* (Oxford: Oxford University Press, 1982).
[6] John Hick, *Faith and Knowledge*, Second Edition (Ithaca: Cornell University Press, 1966), p. 3.
[7] Richard G. Swinburne, *The Coherence of Theism* (Oxford: Oxford University Press, 1977), p. 1.
[8] See, for example, Michael Durrant, *The Logical Status of 'God'* (London: MacMillan, 1973).
[9] For more on this, see the Introduction to *The Concept of God*, edited by Thomas V. Morris (Oxford: Oxford University Press, 1987).
[10] For more details, see Thomas V. Morris, 'Perfect Being Theology,' *Nous* **21** (1987), pp. 19—30.
[11] For more on this, see Thomas V. Morris, *Anselmian Explorations* (Notre Dame: University of Notre Dame Press, 1987), Chapter 1, and George N. Schlesinger, *New Perspectives on Old Time Religion* (Oxford: Oxford University Press, 1988), Chapter 1.
[12] Ibid.
[13] See 'A Theistic Proof of Perfection,' *Sophia* **26** (1987), pp. 31—35, or 'Metaphysical Dependence, Independence, and Perfection,' In *Being and Goodness*, edited by Scott MacDonald (Ithaca, Cornell University Press, 1990).

EPILOGUE

MICHAEL LEVIN

PHILOSOPHICAL ANALYSES: AN EXPLANATION AND DEFENSE

If definitions are what you find in dictionaries, philosophical analyses aren't definitions. The difference runs deeper than concern with use over meaning and the explication of obscure concepts by clearer ones, matters of perhaps less moment to lexicographers than philosophers. Rather, successful philosophical definitions have the ring of empirical discoveries or illuminating theories. Think, for instance, of Hobbes' celebrated definition of "laughter" in terms of "sudden glory," or Russell's of "number" as "maximal set of equinumerous sets." Successful or not, philosophical definitions have been tested like theories since Socrates first vexed his interlocutors with hard cases.

Hard cases provoke refinements which invite harder cases, a dialectical helix which may easily appear to get out of hand. Clauses accumulated to meet outré counterexamples build to a complexity not likely to match anything in the minds of ordinary users of the definienda. This process may be thought to be illustrated by the post-Gettier knowledge literature or the post-Grice meaning literature. Does a child who takes his mother's "Brush!" as an order to brush his teeth really think she uttered "Brush!" to get him to believe she wants him to brush his teeth through his recognition of her intention . . . that he recognize that intention? The material on Goodman's analyses of lawlikeness and simplicity, which is likewise accessible only to specialists, can probably supply further examples of apparent overcomplication.

Appearances notwithstanding, these characteristics are not faults of contemporary philosophical analysis. Current practice can be explained and justified. The explanation I propose here[1] is systematic in that it involves a specific conception of language as a whole and seeks unity in already familiar aspects of philosophical analysis, such as the requirements of extensional and intensional equivalence. It does not, however, seek to resolve any specific definitional dispute and therefore makes no special effort to join any. On the other hand, as metaphilosophy is to many minds that than which nothing duller can be conceived, I seek a stimulating application in the penultimate section by comparing the similarities between philosophical analysis and aesthetic criticism.

J.H. Fetzer, D. Shatz and G. Schlesinger (eds.), *Definitions and Definability: Philosophical Perspectives*, 287–309.

I. WORDS AND THEIR CORRELATES

I assume a broadly behavioristic view of language, according to which an idiolect is a family of associations between utterances and circumstantial cues. However innate may be the capacity to learn such associations,[2] semantics itself is wholly acquired. No-one is born knowing when to assent to "Automobile?" or how to translate Mentalese into its local conventional counterpart. Even if there are innate representations, the problem of determining which public conventions (if any) correspond to one's innate representations can be solved only by conditioned association. I may have [knowledge] in my on-board computer, but that won't tell me whether its translation is "knowledge" or "connaissance" around here. What is ordinarily called the problem of first-language acquisition simply becomes the problem of translating the local lingo into one's real first language. Reversing perspective, innate representations do not by themselves determine which innate representation some given public convention corresponds to. Upon first hearing my neighbor's "knowledge", I can't possibly tell whether to translate it as [knowledge] or [poodle] or any other Mentalese entry. The problem must be solved inductively, by discoveries about conditioning. Really, it is obvious that language is *at least* a system of conditioned habits. I *am* disposed to assent to "Tintoretto?" upon exposure to what I take to be a Tintoretto, whatever else the discharge of that disposition may require. Semantic competence is produced by the reinforcement of assent to and dissent from words in circumstances rendered appropriate by the reinforcement itself.

A more serious issue is the circumscription of cues proper. According to common sense, a few predicates — like "red" — are cued by sensory states, while the rest, like "paleolithic," are associated with non-observable circumstances described in some antecedently understood vocabulary. Assent to non-observational predicates is triggered by observations thought to indicate the presence of the appropriate unobservable circumstance. The sight of a tool elicits assent to "Paleolithic?" because its shape, color and texture is thought to indicate an age of 300,000 years.

If pressed, common sense will agree with Quine that a speaker's verbal behavior at a time t is completely determined by the state of his nervous system at time $t - \delta t$, just as all behavior is immediately caused by the state of the behaving organism's nervous system.[3] Whatever the objective distal cause of the stimuli reaching a speaker and of his present dispositions to respond to such proximal stimuli, the same stimulus will evoke the same verbal response. A half-glimpsed duck and an observationally indistinguishable decoy will both elicit assent to "Duck?" Where Quine departs from common sense is in inferring from this truth that only stimulus meaning is an objective feature of word usage. It is this inference from the determinativeness of neural states with respect to verbal behavior that leads to the inscrutability of reference and the indeterminacy of translation.

Common sense retrieves what it supposes to be objective circumstantial cues by treating a speaker's beliefs as mediating mechanisms. The speaker assents to "Paleolithic?" because he believes (possibly on the basis of yet other beliefs) that a certain appearance indicates an age of 300,000 years. Quine, however, worries about the individuation of beliefs and the division of intersentential associations into the belief-based and the meaning-based. That is why he speaks generically of sentences being animated by stimuli and other sentences. Quine's speaker assents to "Paleolithic?" because he has learned to assent to "300,000 years old?" when shown objects of a certain sort, and to assent to "Paleolithic?" when prepared to assent to "300,000 years old?"

Quine accused Carnap of begging inscrutability questions when Carnap proposed to construe the verbal responses of the hypothetical robot, Karl, in the common-sense idiom of physical objects. Quine would doubtless accuse the common-sense theory of cues of begging the same questions. It is useless to pretend that Quine's challenge to common sense does not exist, but it helps to point out that Quine adheres to common sense more closely than is commonly supposed. The Quinean and common-sense accounts of a speaker's verbal behavior can be correlated, via inference to a word's non-observational cues, from the sentences animated by its positive stimuli. Word *W* has circumstantial meaning *C* in the idiolect of speaker *S* just in case *W* has positive stimulus meaning *P* and *S* is disposed to assent to "*C*?" whenever he is disposed to assent to an observation predicate whose stimulus meaning belongs to *P*. "Paleolithic" is cued by an age of 300,000 years because assent to "300,000 years old?" triggers assent to "Paleolithic?" Non-observational cues are given by description in some antecedently understood vocabulary.

II. IGNORANCE ABOUT CUES

With pangs of guilt about ignoring Quine's scepticism, I turn to the principal lemma in the behaviorist theory of language: conditioning, verbal or otherwise, does not require the subject to know what cue is controlling his response. The subject must of course be able to discriminate the cue, but only in the trivial sense that the cue must influence his behavior. No knowledge is needed.

To cite an illustration of this lemma, most people can be taught to use "art deco" because they can discriminate the art deco style, but few can say why they call an object "art deco." It's not just zigzags, but the average person can do little better when explaining what makes him say "art deco." Most people can tell irony from sarcasm — to cite a second illustration — but few can say how they do it, what quality distinguishes irony from sarcasm. The ability to describe improves with study and reflection, but the typical verbal response is unstudied.

It is useless to look to appearances as criteria, even of words with uniform stimulus meaning, since appearances are not always discernible or describable when discerned. Often the only accessible description of the look of things we incline to call "*W*" is that they are *W*-looking. Art deco objects look distinctive, but someone who professes to use their art-deco-ish look to categorize together the Chrysler Building and a period travel poster may not himself know the objective features determining the look or the components of the visual art-deco-ish gestalt to which he is responding. Art deco objects strike him as similar in an otherwise ineffable way that he has learned to call "art deco."

The lemma is very vividly demonstrated by the ascription of psychological predicates like "aggressive" and "intelligent." Most people are good at dividing their acquaintances into the intelligent and the less intelligent, and most people would divide the same sample roughly the same way, yet most people are bad at articulating what makes them do so. Indeed, a number of intelligent philosophers have concluded from this that "intelligent" is meaningless. (At least, they *say* this is why they find "intelligent" meaningless.) However, these intrasubjective and intersubjective agreements about the treatment of new cases would be inexplicable were there no objective trait of human behavior controlling this usage. It is more plausible to conclude that people can discriminate this trait without being able to say what they are discriminating. (Conceivably, predicates like "intelligence" might be cued by traits — such as being highly educated — about which speakers share a false theory — such as some connection between this trait and a general capacity of the mind. In such a case, however, a predicate like "intelligent" would be perfectly *meaningful* and merely have the empty set for its extension. To use the stock example, "witch" is and always was meaningful — it means "woman who consorts with the devil" — but it was incorrectly thought in old Salem to have a non-null extension.)

As the ironic/sarcastic and intelligence examples suggest, the lemma applies primarily to non-observational verbal cues explained by examples and vague descriptive commentary. Children are taught "shaggy dog story" by exposure to a few anecdotes glossed as "pointless narrative with an irrelevant punchline." They are then expected to assent in future to "Shaggy dog story?" if presented with similar anecdotes. The word "philosophical" is also like that. Even non-philosophers are pretty good at identifying "philosophical" distinctions and a "philosophical" outlook on life, despite the unavailability of any popularly accepted analytic characterization of philosophy.

The point of the lemma is not Putnam's (in his seminal essays on reference of the 1970's), that people can refer to something without knowing what it is. What is shown by the intuitive difference in meaning between "water" and twin-"water," after all, is that meaning — one of whose components is reference — exerts *no* immediate control over verbal behavior. *My* point is that speakers often cannot identify those features that *do* control verbal behavior. Where Putnam's distinction between meaning and stereotype is germane, the association established by conditioning holds between word and stereotype, and the lemma runs that speakers may not know the stereotypic cues of their own words. Where Putnam and Kripke rightly stress the distinction between appearance/stereotype/reference-fixer, on one hand, and reality/reference/rigid-designator on the other, I am stressing the liability of speakers to ignorance about appearances themselves.

If philosophical analysis is the articulation of cues, as I suggest below, it may sometimes be the articulation of stereotypes — a corollary which serves to deepen the felt difference between philosophical analysis and dictionary defining. On the other hand philosophically unclear words such as "know" do not seem like the stereotype-carrying natural-kind terms to which Putnam's analysis most plausibly applies. The only example of a Putnam-like treatment of a philosophical word I can think of offhand is Grice's suggestion that a subject perceives X if

X causes in him a sense-experience in whatever way it may be that objects cause experiences in exemplary cases of perception.[4]

The important point to retain is that language use is, to a significant extent, the application of predicates to new cases on the basis of a sensed but unarticulated similarity to old. Speech is stimulus generalization in action.

III. TROUBLES WITH CUES

Only trouble comes of demanding conscious criteria where cues will do. Although digressive, two illustrations of such trouble help to clarify the principal lemma.

(A) Consider Wittgenstein's paradox about "going on the same way." Wittgenstein saw that communication depends on similarly trained speakers judging new cases similarly. He saw as well that the full use of an expression cannot be fixed by finitely many examples. Since training must be finite, he asked, why doesn't communication constantly break down as speakers' usages diverge in cases not anticipated by training? Recognizing a problem here was a stroke of genius, but what Wittgenstein seems not to have recognized is the adaptiveness for social conspecifics of agreement in their tendency to generalize stimuli of all sorts. Indeed, similar genotypes produce similar phenotypic neural wiring, even among nonsocial conspecifics, so that all conspecifics will tend to "go on the same way" when susceptible of conditioning at all. It is thus only natural that similarly trained human speakers judge new cases similarly. Communication is robust because harmony among speakers — a puzzling stroke of anthropological luck for Wittgenstein and his followers — was pre-established by evolution.

Wittgensteinians also complain that a dispositional theory of language cannot explain correctness — why going on one way rather than another is *right*. What is required for any solution to this problem is a specification of *intrinsic similarity*, by whose lights the stimuli to which speakers generalize training cases may be judged proper extensions. However this problem is to be solved,[5] it is not the problem of de facto mutual comprehension.

(B) Next consider introspection. Materialists think introspective reports like "My back hurts" are cued by neurological states. Initially, speakers characterize these neurological cues in terms of some public circumstances in which they occur, so that the language-learner is taught to use "backache" of what goes on when his back is struck. Although the speaker is conscious of the neurological cue when it occurs, in the sense that it controls his verbal response, he typically knows nothing more about it than that it results from damage to his back. He does not know that it is a neurological cue, for instance.

So far, all is straightforward. Trouble starts with worry about *how* at a later time the speaker detects a similarity between the inner state he is then having and his original paradigm. One common form of relief for this worry about a criterion is the phenomenal property: the speaker noticed the backachey feel of his backache when it first occurred and now uses "backache" of new inner states with the same feel. One push from Platonism about properties, and the unwary are plunged into a netherworld of double aspects and emergent properties. But the worry that conjures phenomenal properties is unreal: speakers need not be guided by observable traits of their nervous systems to reapply psychological predicates. It is enough that there be recurrent states of the nervous system which cause the reapplication of psychological predicates. Physiological states suffice without phenomenal intermediaries to control introspective reports.

Certain forms of eliminativism apparently share the presumption of a criterion. If *all* that is cuing my sensation reports are neurological states — no phenomenal properties needed — belief in inner states is a mistake. If this is the eliminativist argument, it too assumes that reports of psychological states must be true of something more than physical states if they are true of anything. Again, the sensible alternative is to trust the materialist's account of the acquisition of the basic introspective vocabulary and take neurological states as what this vocabulary is about.

IV. VAGUENESS

A cue *C* may fall outside the union of the positive and negative circumstantial meanings of a word *W*, or lie on their mutual border. The former case results in the indeterminacy of application of *W* to *C*, the latter in that special form of indeterminacy called vagueness. Many philosophers take vagueness in particular as a challenge to realism and bivalent logic. The paradox of the heap leads some of them to detect countless inconsistencies in our everyday beliefs, others to fill truth-value gaps by various means. Roy Sorensen, a "gung-ho semantic realist," fears that we face "the heartbreak of Sorites" unless there is some number n such that n-haired men are bald and $n + 1$ haired men thatched.[6] Although again disgressive, it is instructive to explore a naturalistic, cue-oriented approach to vagueness.

If the meaning of "bald" for speaker S at time t is his disposition to assent to or dissent from "Bald?" under a variety of conditions, the thing to do is to graph the readiness (including reaction time) of each speaker S to assent to "Bald?" against the number n of hairs on heads presented to him. (S's sensitivity to hair distribution can be accommodated by graphing his reaction time against a density function; inauthentic consistency-seeking responses can be disrupted by displaying the hair distributions randomly; further refinements are possible.) The result will be a range of values for n over which S assents to "Bald?," a range over which he dissents from "Bald?," and a range within which he won't say. More precisely, we can define a number $u(S, t)$ as the *onset of unclarity* for S at t as the least n at which S's reaction time exceeds that for all preceding m, and S's *range of indeterminacy* as $[r: k < r < j]$, where $k(S, t)$ is the first number beyond which S never assents to "Bald?" and $j(S, t)$ is the first number before which S never dissents from "Bald?" There is of course no a priori guarantee that S's reaction time for $n > u$ will never be shorter than it is for u, or that S will be undecided within his range of indeterminacy, or that $t \neq t' \supset \langle u(S, t'), k(S, t'), j(S, t')\rangle = \langle u(S, t), k(S, t), j(S, t)\rangle$, or that $S \neq S' \supset \langle u(S', t), k(S', t), j(S', t)\rangle = \langle u(S, t), k(S, t), j(S, t)\rangle$. These generalizations may nevertheless be expected to hold as a rule, and they show how usage can be vague without leading to the Sorites or the denial that anyone is bald.

Both Sorensen and the Soritist would probably agree with the descriptive accuracy of such an account, but both might claim to detect an inconsistency in S's verbal behavior as reported. The Soritist would say that S *ought* not hesitate at u because there is no difference between u- and $(u - 1)$-haired men, Sorensen that a difference between u- and $(u - 1)$-haired men sufficient to make S hesitate should make S agree that u-haired men are not bald. The naturalist has answers for both. He informs the Soritist of a real difference between u- and $(u - 1)$-haired men: for subtle neurological reasons of which S among others is unaware, u is precisely the point at which S's tendency to assent to "Bald?" begins to weaken. If meaning is use, the Soritist is wrong abount S's language being inconsistent. The naturalist informs

Sorensen that, by the very character of *S*'s reaction, the changeover at *u* is manifestly too small to get *S* to dissent from "Bald?" "Bald" means what it does mean, not what it should mean. The causal explanation of failure to draw Sorites conclusions thus attributes inconsistent beliefs to no-one.

V. PHILOSOPHICAL ANALYSIS

Let us recall the charge that contemporary philosophical analysis tends toward overcomplexity and psychological unreality. Children use "mean" without thinking Gricean thoughts; traffic officers claim to know the way to the airport without regarding their beliefs as reliably caused, involving no false lemmas, or undefeated by counterpossibilities; statisticians diagnose coincidences without having heard of "projectible predicates."

But the upshot of the lemma is that neither children nor policemen nor statisticians need have these definitions in mind for the definitions in question to describe their use of "mean" or "know" or "coincidence." All that matters is that the cues described by the clauses of these definitions, however complicated, in fact control (or "make an impression on") the speakers. In short, philosophical definitions are descriptions of verbal cues. They attempt to articulate the ill-understood factors that elicit speaker response. They seek the causes of verbal behavior of which speakers themselves may be unaware. Philosophical analyses share the a priori feel of dictionary definitions because they concern conditions for word usage rather than the behavior of the reference of words. But they are empirical and exploratory in a way that analytic definitions cannot be.

To put an example on the table, let us assume the causal theory of perception is correct. Suppose young Eric is told, "Mark must have seen the book" when Eric sees a book in front of Mark's eyes with the lights on, and "That was a bell you heard" when a bell rings in Eric's vicinity. Suppose Eric is also told that no-one has yet seen the surface of unexplored Pluto, and that nobody will ever see a troll because trolls do not exist. Now, despite having been disposed by such training to apply perceptions words in cases he finds similar to the first two, and to withhold perception words in cases he finds similar to the second two, Eric may not himself know what these similarities are. It may not even occur to him that he doesn't know. It may take an outsider to discern, by a subtle sifting of cases, that Eric is disposed to use perception words just when causal processes directly produce experiences. Should Eric eventually think about the matter, he may embrace the proffered analysis himself, discovering with a shock of recognition that he uses perception words in just this way. Note one specific detail in this little fable: young Eric never hears an ostensive definition of "perception." Nobody ever shows and tells him "This sort of thing is called 'perception'." Children learn words in context.

I am *not* proposing psychological realizability — or psychological realizability in children — as a novel adequacy condition on defintions. It is in fact self-evident that children must be up to being conditioned to the analysantiae of terms they actually use. What this truism entails, rather, is that psychology may be read off semantics. If a proposed explication of "know" (for instance) appears on independent grounds to capture the cues to which adult speakers are responding — and, so far as we can tell, children use "know" as adults do — then children must be psychophysiologically capable of responding to these same cues. If a Gricean theory of meaning is otherwise plausible, children who use "mean" as adults do must notice and *a fortiori* be capable of noticing Gricean nested intentions. Children need not have the faintest idea of the existence of Gricean intentions, but their sensitivities must be such as Gricean intentions can control. The assumption of developmental constancy of meaning is, for the most part, quite plausible. It is possible but unlikely that puerile usage is controlled by different cues which produce like response in like circumstances. Children admittedly use words idiosyncratically, especially when words carry important practical consequences. A child may call one of his own palpably deliberate acts "accidental" because, so far as he can tell, "accidental" is the adult world's prologue to exoneration. These cases are easily spotted, however.

In seeing the ages at which children master words as signposts of development, I am not claiming to deduce conclusions of empirical substance from premises about mere words. The point I am making can be reformulated quite straightforwardly in the idiom of concepts. As the Piaget school has shown, the mastery of a number of general basic concepts — number, geodesic, conservation, equality, fairness — is age-dependent, a dependence which in turn expresses an innate program presumably biological in nature. My idea is Piaget's, minus the intermediary of concepts, put in terms of the (concept-expressing) words children are called upon to master.

I have so far conceded the complexity of philosophical definitions, and I cannot very well deny this complexity without threatening the theory I have proposed to accommodate it. However, this complexity should not be overstated. Many simple things — verbal cues emphatically included — sound abstruse because their descriptions are. Take Nozick's definition of knowledge,[7] according to which *A* knows *p* only if *A* does not believe *p* in the first doxastic alternative to *p*, i.e. the nearest possible worlds in which $-p$. This sounds incomprehensible to all but professional philosophers, but only because of the presence of terms of art like "doxastic alternative." What Nozick is getting at is quite straightforward: If Jones really knows it is 12:00 and is not right by coincidence, he would not have believed it was 12:00 had it not been noon — had it been 11:00, for instance, the clock he looked at would have said 11:00. Assuming it to be otherwise adequate, this idea is simple enough for Nozick to be credited with explicating the *ordinary* concept of knowledge. Much the same can be said for Goodman's explications of "cotenable" and "about," Grice's analysis of nonnatural meaning, and other allegedly exotic specimens in the philosophical zoo.

Precision wars with clarity, said Peirce; terminological complexity tends to disguise deep-seated intuitive simplicity when perfect accuracy is demanded. This, it is my impression, is a combination encountered with especial frequency in law and mathematics. While the idea behind "heirloom" is natural enough, its legal definition occupies several pages of Trollope's *The Eustace Diamonds*. Probably no idea is more immediate than that of smooth curvature, but its precise definition requires the definition of continuity, its contrast with differentiability, and theorems which few college students ever master. Or, finally, take the simple idea of a quantity x periodically changing signs, which is captured by the term $i^{2n}x$, i being $\sqrt{-1}$. It generally takes a moment to realize that this term is $1x = x$ when $n = 0$, $-1x = -x$ when n is 1, $1x = x$ when n is 2, and so on. Since the dip of a wave can be expressed as a sign-change of its amplitude, branches of physics dealing with wave phenomena, like quantum mechanics, find it useful to employ functions in i. Much of the forbidding appearance of this material is notational in origin.

VI. PHILOSOPHICAL ANALYSIS (CONTINUED)

J. L. Mackie was the first to observe that analysis construed as the display of cues could resolve the paradox of analysis. A display of cues is genuinely informative. It is no mere reminder or explication of what the speaker knows implicitly, since the speaker may not know in any sense what is cuing him. A characterization of cues may be derivable from thought-experiments the subject can perform — as I will argue shortly — but the results of experiments are paradigmatically newsy.[8]

The standard requirement that definitions preserve extensional equivalence becomes, in the present context, the requirement that, for circumstance *C* to cue word *W*, *C* must be present when and only when *S* assents to "*W*?" (Here I am talking about something like total cue.) But extensional equivalence, as is well known, is insufficient for analysis or definition generally. If all and only spheres were red, it would remain necessary to distinguish words cued by sphericity from words cued by redness. In our world the observable features of combustion occur when and only when oxygen atoms donate electrons, but electron transfer does not cue "fire" as it is ordinarily used. We wish to allow *C* but not *C′* to cue a word *W* even when $(x)\ (Cx \equiv C'x)$.

Intuitively, this means that *S* would assent to "*W*?" in the presence of *C* without *C′* but not in the presence of *C′* without *C*. Philosophers avoid counterfactuals as if they were radioactive, and they should perhaps rightly avoid treating counterfactuality as a primitive notion; but the counterfactuality of the criterion just stated is in itself unproblematic. Determining the range of triggers of verbal dispositions is precisely like determining the range of triggers of the solubility of sugar, which outrun the conditions to which sugar has been or (presumably) will be exposed. When solubility and verbal dispositions are identified with crystalline and neural microstructure respectively, both dispositions can be extrapolated from the behavior of objects exposed to a test condition to structurally similar unexposed objects. A sample of sugar would dissolve in a chemical in which structurally similar samples of sugar have dissolved, and a speaker *S* would assent to "*W*?" in circumstances under which neurologically similar speakers have so far assented. In this respect, counterfactuals about assent are as tame as counterfactuals about solubility.

The two cases part company, unfortunately, because the antecedents of counterfactuals about assent may flagrantly violate natural laws. If *T* occurs just when *T′* does, science is typically uninterested in what would happen given *T* without *T′*. Chemists do not worry whether sugar would dissolve in 100° C liquid hydrogen, since hydrogen is gaseous at 100° C. But just such probes of assent are necessary to differentiate verbal cues. We must make empirical sense of speakers' dispositions to assent to "Fire?", given flames without electron donation, even if such flames could not occur.

To make empirical sense of such a probe, think of an introspective sugar cube concerned not with its likely fate if immersed in a variety of fluids but with its likely feelings upon immersion. Like us, the sugar cube can evoke feelings in itself by imagining itself in various situations. These feelings are fallible but useful indicators of how it would feel in the situations it is imagining, just as the feelings a man evokes in himself by imagining himself winning the lottery are fallible but useful indicators of how he would feel if he actually won. Like us, furthermore, the sugar cube can sometimes evoke feelings in itself by imagining what it may know to be impossible. It can imagine how it would feel if poised for a bath in 100° C liquid hydrogen, just as I can imagine how I would feel if asked to pilot the first faster-than-light spaceship. Such thought-experiments survive cotenability problems and ask no special status for impossibilia.

If in point of psychological fact the thought of 100° C does not exclude the thought of liquidity, the sugar cube can introspect its response to the *thought* of warm liquid hydrogen. Counterfactuals about the behavior of sugar in 100° C liquid hydrogen may be ill-formed, since warm hydrogen is gaseous; but statements about the cube's possible feelings in that situation are quite clear: they are categorical statements about how the cube does feel when imagining the situation. The disposition thus probed by imagination is not solubility, which can be probed only by realizable environments, but the emotional character of the cube itself. The logical form of descriptions of this thought-experiment require only that the monadic predicate "is a warm-liquid-hydrogen-imaginer" be ascribed to the cube, the result of the experiment being the psychological states consequent upon this predicate's being satisfied.

Thought-experiments about the impossible, like those about the possible but unactual, yield empirical information about the cube's attitudes. The cube need not know before imagining them how it would feel in unrealized circumstances. Imagining oneself in a novel circumstances is like prodding oneself with a novel instrument. The imaginative evocation of feeling, while reliable, is emphatically fallible: the cube may not react as imagination led him to suppose he would when actually dunked in a solvent he had previously only thought about. This objective control over imagined reactions is absent in the case of impossibilities, but thought experiments about impossibilities produce information about the subject's mind nonetheless, by producing categoricals about his subjective responses. These thought-experiments probe the same microstructure probed by thought-experiments about possibilities and, indeed, exposure to actualities.

A speaker may be likened to our reflective sugar cube, with respect to his verbal rather than attitudinal dispositions. He can discover what he would say in unencountered situations by imagining himself in them. Third parties may ask him to think about these situations, or he may conjure them up as a do-it-yourself exercise. Prodding internal multi-tracked verbal dispositions produces genuine information about what the speaker would say if asked to judge hiterto-unconsidered possibilities or impossibilities. All that matters is that the (im)possibilities be psychologically vivid. A speaker can probe his own readiness to use "see" by asking himself if he would say a subject "saw" a lamp if Martians diverted the light from the lamp while telephathically transmitting appropriate visual experiences in the subject. It doesn't matter for the reliability of this prediction that there can't be telepathic Martians, since what is being tested are the saliencies which excite the speaker's verbal dispositions. Should the speaker feel disinclined to use "see," he has discovered that the directness of the causal path from object to experience is one of his cues. Statements about what the speaker would say in response to impossible Martians can be exchanged for categorical statements about what he does say when imagining them. Again, the logical form of such experiments requires ascription of the predicate "is a telepathic Martian imaginer" and a reading of verbal responses consequent upon his satisfying this predicate.

The need to vary factors in imagination further complicates philosophical analysis. Isolating the effect on verbal behavior of a factor entwined in laws may demand stories as fantastic as those in the Gettier or personal identity literature. But appeal to variation beyond the possible also deepens the sense in which philosophical analyses *discover* tracks of verbal dispositions. Those who have never considered Gettierological bizarries are ignorant of something about their use of otherwise familiar epistemic terms. That a man's verbal dispositions reside in him as microstructure no more assures his access to everything they will do than his stomach's residence inside him assure his knowledge of its reaction to ingestible substances he has never tried and perhaps never heard of. Take seriously the idea of language as verbal disposition, and the idea of discovering what you mean becomes entirely natural.

Since thought-experiments reveal reactions to imaginings of situations rather than situations themselves, such experiments do not infallibly predict what one will say about real situations. Perhaps I say *now* that a person replica isn't a "person," but I would in fact find myself calling them persons if they ever came around. Counternomologicals and counter-what-are-not-known-not-to-be-nomologicals omit a lot and, the more telegraphic they are, the more they omit. This is perhaps another source of dissatisfaction with philosophical test cases. Paragraphs about amoebic fission may evoke different responses to "Person?" than do novels about the same subject. HAL in *2001* is a better advertisement for machine consciousness than a journal article because of the superior vividness of moving images. Yet, while this consideration points up the need to stay alert for all relevant details, it shows no inherent instability in cue-elicitation experiments. If you find a speaker, possibly yourself, who thinks he uses *W* when and only when *C*, but discovers himself willing to assent to "W?" in a concocted *C*-less case or to dissent from "*W*?" in a concocted case involving *C*, another trip to the drawing board is in order.

This view of probing linguistic intuition may seem oddly passive and may seem to ignore the active aspect of the rendering of linguistic verdicts. One *thinks* about Gettier cases when first presented with them; one doesn't just sit back and wait for something to happen in the way of a verdict. In fact, however, I think this *is* close to what really happens. The rendering of such judgements *is* essentially passive, and we do just observe — and, if so inclined, report — our responses. Certainly I don't *decide* whether to think Smith knows that a man in his office owns a Ford; how the case strikes me isn't up to me. Whether to tell you how the case strikes me is up to me, but that is another matter.

I am so confident of the involuntariness of linguistic judgements that I hereby offer $10 to anybody who changes his mind at will about Jones' epistemic state. If you don't feel an impulse to claim the bounty, you recognize that you can't help thinking whatever you do about Smith's epistemic state; and if you can't help thinking what you do, thinking what you do isn't within the ambit of your will. Reports of what you thought when you first heard Gettier examples describe something that happened to, or in, you. Linguistic verdicts do admittedly sometimes take time, thought and above all review of salient gedanken-features, but what all that mulling over seems to amount to is the parading of salient features before the mind to let them impress us. Nor do we even decide which features to attend to: as the idiom has it, they occur to us. Saliency, too, is something that happens. It is admittedly difficult to find appropriate language for the strange blend of activity and passivity involved in thinking, perhaps because of the persistent illusion that thought is controlled by a will.

Loose ends. It has been suggested to me that philosophical definitions have two components, a core of clauses capturing the intuitive content of the definiendum and a periphery of clauses to ward off counterexamples. Precedent for this suggestion are the two tiers of axioms for set theory. Some serve the prestigious function of generating mathematics, while others, such as Regularity in the Zermelo-Fraenkel system, exist only to block paradox. This distinction is untenable, however, for *ZF* or systems generally. A definitional clause designed solely to block countercases, with no other motivation or independent plausibility, would be transparently ad hoc and thereby a failure. If Regularity only stopped the paradoxes and captured no other set intuitions, it would but restate the paradoxes rather than solve them. By the same token, telepathic Martians remain to be dealt with if a definition of "perceive" fends them off with a clause that amounts to disqualifying telepathic Martians. No clause is more equal than any other in a proper definition.

On the other hand, there is no a priori reason why philosophically important words should be everywhere defined. There is indeed much presumptive evidence that some are vague. "Person," for instance, was invented and transgenerationally taught on the basis of paradigms, namely, adult human beings. If people don't know quite what to say about fetuses, this may be because fetuses strike them as somewhat like adult humans and somewhat different from adult humans. (If adult humans are part of a stereotype that fixes an underlying scientific essence, the stereotype itself is vague and so is less apt to have fixed a precise scientific essence.) One rebels at their being no objective totally defined criterion for personhood, because personhood decisions are important. But this may show only that some practically important matters have to be decided by stipulation.

VII. AESTHETIC CRITICISM

Verbal responses are not the only puzzling manifestations of our inner lives. The wide range of responses to works of art also needs explaining, and I think my account of philosophical analysis can be extended to aesthetic criticism. I suggest that one important aim of criticism is the explanation of aesthetic responses in terms of factors that spectators notice but are not aware of noticing.

Some account of what critics are up to is surely needed, since so many current conceptions are unpersuasive. It is widely thought that the purpose of criticism is evaluation, but, while evaluation is certainly one dimension of criticism, it is far from the most central. "Service" reviews in popular periodicals help readers decide which movies to see and books to buy, but such practical recommendations are consequences of critical insight, not critical insight itself. Once he has mentioned the looseness of plot in a Robert Ludlum novel and the lurid overwriting, the critic need not add in so many words that the novel fails. Certainly, the critic adds nothing to what he has already stated if he does append those words. It is fun to rank the famous — "The three greatest lyric poets are . . ." — but that is little more than a parlor game. As Nelson Goodman has urged, merit is one of the critic's means, to the end of understanding:

> If a connoisseur tells me that one of two Cycladic figures that seem to me almost indistinguishable is much finer than the other, this inspires me to look for and may help to find the significant differences between the two.[9]

At the same time, I am not persuaded by Goodman's idea that sensitization to art enhances understanding of the world. *Macbeth* has not clarified ambition for me, but acquaintance with human folly over the years has clarified *Macbeth*. Nor does criticism aim at enhancing appreciation of art itself, if only because criticism intensifies experience too haphazardly. It is interesting to learn that the unstable postures of Caillebotte's *Paris, A Rainy Day* reflect the growing influence of photography, and one may first notice this detail upon learning so, but, in my experience, such information does not make the painting look any better. Analytical insight into musical composition is even less apt to make the music sound better.

Hence my proposal that one major goal of criticism, perhaps *the* major goal, is increased understanding of the beholder's response to artworks. Just as philosophical analysis identifies the causes of verbal response, criticism identifies and articulates those features of artworks which produce the range of emotions associated with art. Raising these features to prominence need not intensify the beholder's primary experience, but it serves to explain why a work evokes the primary experience it does. Just as people cannot say why they call a poster or item of jewelry "art deco," they cannot say why one melody rather than another haunts them, or why the Flatiron Building is more impressive than much larger International Style buildings.

Saying why is a job for critics. The greater variety of responses the critic must consider tends to obscure the methodological similarities between analysis and criticism. Determining why *Hamlet* is mysterious is a less sharply defined if not more difficult task than determining what makes people use "refer" as they do in Twin-Earth cases. For the same reason, criticism employs a broader range of psychological, historical and scientific techniques, some of which may seem pedantic but all of which can become unexpectedly pertinent. Harry Levin, for instance, has counted seventeen occurrences of the word "question" in *Hamlet* and seventy question marks in the 332 lines of the graveyard scene, two stylistic factors which "contribute to the general atmosphere of uncertainty" surrounding *Hamlet.* Levin's observation helps explain the uncertainty everybody feels.

Or consider the consensus view that *Othello* is Shakespeare's most tightly plotted play. In addition to the obvious reasons, there is the intriguing possibility that the catastrophe seems inevitable because Othello is *right* to fear Desdemona's attraction to Cassio. Othello realizes "She loved me for the danger's I had pass'd, and I loved her that she did pity me." Aware at some level of the impermanence of such infatuation, the audience dates the tragedy not from the accident of Othello's falling under Iago's influence, but the unwise marriage itself. To the extent that this explanation is correct, it brings to consciousness an aspect of the play that affects audiences without their realize it.

Just as an audience need not know what it is responding to and a language teacher need not know the cues of a word he is transmitting by paradigms, an artist need not be aware of the features he inserts in his work nor why he has inserted them. Shakespeare might not have realized why *Othello* would have been less gripping if its protagonist had been a nobleman. If asked why he made Othello a Moorish soldier, Shakespeare might have replied in all honesty, "I don't know; it just seemed right." His genius lay in having the right plot device occur to him and selecting it when it did occur. Commentators since Plato have wondered that poets themselves don't know how they do what they do; thus T. S. Eliot's paradox that "the bad poet is usually unconscious when he ought to be conscious, and conscious when he ought to be unconscious." It may take a philosopher to explain ordinary usage to ordinary users, and a critic to explain the author's judgement to the author himself.

The elucidation of unnoticed causes may also be an aspect of Keats' "negative capability," the capacity to be "in uncertainties, mysteries, doubts, without any irritable reaching after fact and reason." Perhaps Keats had in mind, in part, the power to discern an emotionally significant gesture or image or situation without knowing why it is significant. The movie *Colors*, about black youth gangs in Los Angeles, was simultaneously condemned for racism and for glorifying black violence. Responding to these contradictory accusations, its director, Dennis Hopper, told a newspaper, "I just pointed my camera and said 'Observe.'" Critics were judging what Hopper showed them; Hopper's own contribution ended with his realization that gang violence is riveting and his very possibly instinctive sense of what would heighten the reality he wished to present. His film completed, the director becomes one more spectator of a work whose power may mystify him as thoroughly as anyone else.

In minimizing the author's *conscious* awareness of what he is putting into his work, I am not siding with critics who dismiss authorial intention as irrelevant to understanding. Parallel causal constraints are implicit in my accounts of analysis and criticism. Just as a cue for a word *W* must actually (even if unconsciously) contribute to assent to "*W*?", a feature of a work of art helps explain its production only if the author was responding (even if unconsciously) to that feature. The constraint on analysis might seem too strong because a causally inert feature coincident with *S*'s use of *W* is difficult to imagine. However it appears necessary in mathematical contexts. The cue for "straight line" is a certain visual gestalt, even though the presence of that visual gestalt is intensionally equivalent to the metric tensor going to 0.

The corresponding constraint on criticism is easily established by a thought-experiment. Suppose Alexander Pope had been carefully guarded from birth so that he never read and never even heard of the *Iliad*, the *Odyssey* or any other epic poem. Somehow, he happened to produce *The Rape of the Lock*. I think, intuitively, in that case *The Rape of the Lock* would not be a satire of epic poetry. The most that could be said is that *The Rape of the Lock* reminded the reader of other poems, and that it would have been a good satire had its author been caused to write it by acquaintance with other epic poems. A factor relevant to understanding a work must have, consciously or unconsciously, affected the mind of its creator. Indeed, since consciously intending the presence of a feature is a central example of a feature affecting a mind, the intentions of creators count heavily in understanding their works. If Shakespeare stipulated the reality of the ghost, the ghost is real. If historical research suggests that Shakespeare and his audience believed in ghosts, there is a strong presumption that the ghost must be treated as real. It does not matter whether we can see the ghost as a projection of Hamlet's doubts about Claudius or a hologram displayed by Claudius to drive his nephew crazy. These would be *mis*interpretations, and it is tolerance for such misinterpretations that brings the "new critics" and their even newer deconstructionist followers into disfavor.

VIII. CONCLUDING REMARKS

My account of philosophical defining applies in principle to any word whatever. It thus fails to explain why the cues of "know," "perceive," "theory," "person," "choose" and a few dozen other expressions are in special need of elucidation.

A complete answer would require an account of philosophy itself, for which the reader will have to inquire elsewhere, but what I have said about analysis is certainly *consistent* with some familiar metaphilosophical theories. Wittgenstein, for instance, saw philosophy as "the bewitchment of the intelligence by means of language." This is said to occur when overemphasis on one aspect of an expression's use creates an inappropriate "model" for understanding its referent, or when a special use for an expression is played off equivocally against its ordinary use. Emphasizing the nominative character of sensation-words creates the illusion of mental entities (to illustrate the first idea), and calling expectations about ordinary objects "hypotheses" reinforces scepticism (to illustrate the second).

This bewitchment creates the conflict between the meaning of a word and its presumed denotation that is characteristic of many philosophical problems. The predicate "*A* knows that *p*" appears to be true of many (ordered pairs of) things, while the definitional requirement that knowledge exclude all doubt appears to prevent its ever being true of anything. As conflicts of this sort stem from a standard implicit in the predicate which its putative extension seems unable to meet, it is natural in such cases to examine more closely just what we are trying to convey when using a problematic expression. Whatever we think we mean, perhaps what we do mean is consistent with the word's having the extension it seems to have. Perhaps the daunting standard isn't there; maybe "know" doesn't require the exclusion of Cartesian demons. Certainly, many words of philosophical interest are involved with some sceptical problem.

Yet this diagnosis of philosophy is overly narrow. The present account of analysis suggests another one: it is characteristic of philosophical problems that *all the facts are open to view yet we are still mystified*. This characterization covers puzzlements like the one I mentioned at the outset, namely "what is humor?" Everyone has laughed at and told jokes, yet nobody seems to know what a joke is or what makes one funny. It also covers the main questions of technical academic philosophy like scepticism. Everyone uses "know" constantly, yet there is no consensus about its meaning or even the nonemptiness of its denotation. Finally, this characterization also covers those semi-articulate expressions of wonder which inform the popular conception of philosophy. Why is there a universe? Is anything right or wrong? Does human existence have meaning? Nothing about human life is *hidden* — at any rate, it is characteristic of these large questions that (we feel) no currently unknown datum about the universe or human life would resolve them.

There is much to be said for the Kantian diagnosis that such puzzles overgeneralize questions appropriate within experience, but they also seem to me to resemble those maddening "brain teaser" problems. You are told all you need to know about the eating habits of goats, wolves (and cabbages), but how do you get them all across the river? The difficulty isn't quite like that of understanding hard mathematics, which involves seeing an unobvious implication from premises to conclusion. Analogies can only be pressed so far, but the theme of bafflement without ignorance — of answers available yet unavailable — does appear to run through philosophical analysis and philosophical problems generally.

Department of Philosophy
City College of New York

NOTES

[1] This proposal was first made in J. L. Mackie, *Truth Probability and Paradox* (Oxford U.P.: Oxford, 1973). I develop it in too-compressed form in Michael Levin, *Metaphysics and the Mind-Body Problem* (Oxford U.P.: Oxford, 1979), pp. 11—44.

[2] Language-learning meets the standard ethological criteria for innateness: irreversible appearance in virtually all conspecifics, with strength of emission depending on age, in all environments in which the species evolved. Virtually all two-year-olds learn to speak if exposed to sounds to mimic, and no child regresses to babbling after first producing Victorian prose. Certain recognitional capacities ('predator') and perhaps some features of grammar may also be innate. For more on innateness, see Michael Levin, *Feminism and Freedom* (Transaction: New Brunswick, N.J., 1987); pp. 55—66, and James Gould, *Ethology* (Harper and Row: NY, 1982).

[3] "It is important to think of what prompts the native's assent to 'Gavagai?' as stimulations and not rabbits. Stimulation can remain the same though the rabbit be supplanted by a counterfeit. Conversely, the stimulation can vary in its power to prompt assent to 'Gavagai' because of variations in angle, lighting and color contrast, though the rabbit remain the same. In experimentally equating the uses of 'Gavagai' and 'Rabbit' it is stimulations that must be made to match, not animals" (*Word and Object* [MIT Press: Cambridge, Mass., 1960], p. 31). The closest ethological analogues of predicates whose meanings are their stimulus meanings — the observation predicates of the animal world — are fixed action patterns, stereotyped sequences of muscle movement triggered by definite sensory stimuli in accordance with an innate program. A greylag gosling will peck at a red dot whether or not it is on an adult female's beak. However, most animal responses, like most human responses, are mediated by learning. Furthermore, there is a clear sense in which even the presence of fixed action patterns are to be explained by distal objective causes. Thus, goslings peck at red dots because — mother geese having such dots where they keep their food — such behavior is adaptive in the wild. Both sorts of explanation parallel possible explanations of the verbal responses of speakers in terms of the distal causes of proximal sensory stimulation. It is thus unclear why distal objective causes should be excluded from the objective portion of meaning, even if the immediate sensory causes of verbal behavior can be produced without the actual distal cause.

[4] H. P. Grice, 'The Causal Theory of Perception,' in R. Swartz, ed., *Perceiving, Sensing, and Knowing* (Doubleday: NY, 1965), p. 463.

[5] Kripke formulates the problem in *Wittgenstein on Private Language and Following a Rule* (Harvard University Press: Cambridge, Mass., 1982): Should *S* call both an adult male human being and a rock glimpsed in shadow a "man," the dispositional theory seems committed to including both in the positive circumstantial meaning of "man." His

calling the rock a "man" becomes an indication, merely, of idiosyncratic usage, not a mistake. Quine replies that *S*'s having been taught a word with so complex a meaning can be excluded by the invocation of "hypotheses regarding plausible processes" (Quine and Ullian, *The Web of Belief*, 2nd ed. [Random House: NY, 1978], pp. 53—54). Quine also suggests — I have been unable to track down the reference — that *S* made a mistake by his own lights because he would not have assented to "Man?" had he had a clearer view of the rock. These criteria together may seem to construe correctness as too dependent on what seems correct to the speaker, but on close inspection they give intuitively satisfying results. If *S* persists in calling the rock a man under better lighting conditions, we would no doubt conclude that, somehow, he has picked up non-standard associations for the phoneme "man" and was not making a mistake.

[6] See Roy Sorensen, *Blindspots* (Oxford University Press: Oxford, 1988), pp. 216—252, and references therein.

[7] See Robert Nozick, *Philosophical Explanation* (Harvard University Press: Cambridge, Mass., 1981), pp. 172—247.

[8] For a Platonistic attack on the paradox, see George Bealer, *Quality and Concept* (Oxford U.P.: Oxford, 1982), pp. 74—77, 171. According to Bealer, someone who knows there is a circle in front of him can be said to know there is a set of points equidistant from a given point in front of him (since circles are sets of points, etc.), if 'set of points [etc.]' denotes the unanalyzed concept of a set of points, etc. He does not however know there is such a set of points in front of him, if by 'set of points [etc.]' we refer to the fully analyzed concept of such a set of points. For a strikingly similar view from a strictly naturalistic perspective, see Fred Dretske, *Knowledge and the Flow of Information* (MIT Press: Cambridge, Mass.), pp. 215ff.

[9] Nelson Goodman, 'Merit as Means,' in *Problems and Projects* (Bobbs-Merrill: Indianapolis, Indiana; 1972), pp. 120—121.

INDEX OF NAMES

Abelson, R. 79, 80, 81, 108
Ackerman, D. 64, 68
Adams, M. 283
Alice 23
Alston, W. 202
Andreski, S. 108
Andrews 254, 255
Anscombe, G. E. M. 109
Anselm, St. 278
Aristotle 3, 4, 5, 72, 148, 241, 281
Armstrong, D. 7, 75, 76, 107, 130
Asquith, P. 182, 183
Audi, R. 283
Austin, J. L. 9, 17, 65, 68, 139, 157
Axinn, D. 108
Axinn, S. 108

Baker, 254, 255
Baker, G. 266
Barrett, R. 237
Barwise, J. 147, 157, 183
Bayles, M. 13, 16, 253
Becker, L. 87, 108
Benacerraf, P. 202
Bennett, J. 129
Bentham, J. 262
Berger, A. 11, 12, 16, 223
Bergmann, G. 129
Beth, E. W. 9, 10, 142, 157, 165, 173, 174, 176, 177, 181, 183
Bill 43, 57, 58
Birmingham, R. 266
Black, M. 109
Bogdan, R. 159
Boltzmann, L. 242
Bonjour, L. 11, 202, 206, 219
Buchler, J. 68
Butts, R. 158

Caillebotte 303
Callaway, H. 130
Cantor, G. 154
Cargile, J. 6, 15, 21, 47, 50
Carnap, R. 5, 131, 145, 158, 192, 193, 289
Causey, R. 183
Chang, C. 142, 143, 158, 183
Charles 31
Chomsky, N. 89, 129
Clark 262
Claudius 306
Clausius, R. 242
Clay, M. 182, 183, 202
Cohen, M. 9, 17, 108, 136, 137, 138, 154, 156, 157, 158
Comte, A. 136
Cooper, J. 220
Craig, W. 171, 183
Cresswell, M. J. 131, 150, 158
Czermak, J. 158

Davidson, D. 12, 59, 129, 240, 250
Davies, B. 283
Davis 264
Dennett, D. 91, 109
Descartes, R. 84, 307
Desdemona 304
Downey, J. P. 27
Dretske, F. 201, 202, 309
Dreyfus, H. 7, 78, 108
Duhem, P. 77, 108, 116
Dummett, M. 31, 32, 50
Durrant, M. 283
Dworkin, R. 266

Ed 190, 191
Eddington, A. 95
Einstein, A. 27, 28, 178
Eliot, T. S. 305
Elizabeth, H. R. H. 75

Eric 295, 296
Everhart 264

Feigl, H. 107, 266
Fetzer, J. H. ix, 3, 6, 16, 51, 52, 55, 56, 59, 63, 66, 67, 68
Fine, A. 108, 183
Fine, K. 82
Flew, A. 266
Fodor, J. 7, 53, 54, 68, 78, 90, 108
Foley, R. 11, 206, 219
Forester, C. S. 92
Frege, G. 41, 154, 155, 158
Friedman, M. 202

Galileo, G. 230, 231, 232, 233
Gallie, W. B. 92, 93, 109
Gettier, E. 187, 191, 201, 287, 300, 301
Gibson, R. 237
Giedymin, J. 183
Gilbert, M. 131
Glover, J. 88, 108
God 14, 15, 31, 190, 269, 270, 271, 272, 273, 274, 275, 276, 277, 278, 279, 280, 281, 282
Goldman, A. 11, 189, 200, 201, 202, 206, 219
Goodman, N. 7, 71, 86, 108, 131, 145, 158, 208, 209, 213, 214, 287, 297, 303, 309
Gould, J. 308
Green, M. 88, 108
Greenawalt, K. 266
Grice, H. P. 112, 129, 287, 291, 296, 308
Griliches, Z. 182

Hacker, P. 266
Hallden, S. 77
Hamlet 306
Hanson, N. R. 119
Hansson, B. 13, 247, 250
Hardy, E. 282
Hare, R. M. 266
Harman, G. 112, 114, 115, 120, 121, 123, 124, 130, 131, 202, 219, 237
Harris, S. 183
Hart, H. L. A. 259, 261, 262, 266
Hegel, G. W. F. 43
Heisenberg, W. 106, 180
Hempel, C. G. 4, 5, 16, 17, 90, 91, 109, 156, 157, 158, 177
Henkin, L. 182
Hick, J. 275, 276, 283
Hintikka, J. 9, 10, 16, 129, 143, 144, 146, 155, 156, 157, 158, 159, 161, 182, 183
Hiram 41
Hobbes, T. 121, 287
Hopper, D. 305
Horace 41
Hornblower, H. 92
Hsiao, C. 182
Hume, D. 78, 215

Iago 304
Ingram, P. 109
Intriligator, M. 182

James, W. 92
Jamieson, D. 131
Jammer, M. 182
Jean 75
Joe 191, 200
John 94
Johnston, M. 88, 89, 108
Jones, 279, 301

Kamp, H. 83
Kant, I. 307
Kaplan, M. 201
Karl 289
Katz, J. J. 90, 108, 129
Kaufman, G. 283
Kearn, J. 105, 109
Keats, J. 305
Keisler, H. 142, 158
Kepler, J. 71
Kim, J. 12, 196, 202, 240, 250
Kitcher, P. 182
Klein, P. 202
Krausz, M. 108
Kretzmann, N. 276

Kripke, S. 238, 291, 308
Kueker, K. 143, 144, 158
Kyburg, H. 183

Lehrer, K. 7, 71, 95, 97, 109, 182, 183, 202
Leibniz, G. 27
Leplin, J. 183
Levi, I. 201
Levin, H. 304
Levin, M. 15, 287, 308
Lewis, D. 8, 13, 108, 115, 118, 127, 128, 130, 131, 145, 158, 245, 246, 250
Lewy, C. 37, 50
Linus 125
Llewellyn, K. 267
Lucy 125
Ludlum, R. 303
Luper-Foy, S. 11, 16, 203
Lycan, W. 7, 8, 16, 71, 72, 84, 108, 111

MacCallum, G. 267
MacDonald, S. 283
Mach, E. 9, 10, 164, 182
MacIntyre, A. 219
Mackie, J. L. 297, 308
Makkai, M. 143, 158, 183
Makowski, J. 143, 158
Malcolm, N. 91, 92, 93, 94, 109
Mark 295
Marsha 200
Martin, R. L. 108
Mary 94
Matthews, B. 83, 108
Maxwell, G. 107, 266
Maxwell, J. C. 242
McKinsey, C. 182
Mill, J. S. 241
Miller, R. 282
Moles, R. 266
Moore, G. E. 8, 94, 115
Moore, M. 257, 258, 266
Mooris, T. V. 14, 15, 16, 269, 283
Morse, J. 254
Moses 270
Munitz, M. 108
Munzer, S. 266

Nagel, E. 9, 17, 79, 108, 136, 137, 138, 154, 156, 157, 158, 237
Nazianzus, G. 270
Nerlich, G. 130
Newton, I. 27, 28, 164, 178, 207
Nietzsche, F. 88, 97, 108
Niiniluoto, I. 159, 248, 250
Nowell-Smith, P. 250
Nozick, R. 15, 191, 202, 297, 309

Objibway, F. 253
Othello 304

Parfit, D. 84, 85, 108
Pearce, D. 149, 158, 183
Peirce, C. S. 7, 16, 55, 62, 68
Penny 25
Piaget, J. 296
Pindar 274
Plantinga, A. 270
Plato 79, 188, 281, 305
Platts, M. 108
Pope, A. 305
Priest, G. 109
Przelecki, M. 145, 154, 155, 158
Putnam, H. 15, 27, 28, 29, 32, 50, 120, 131, 266, 291

Quine, W. V. O. 7, 8, 12, 13, 15, 16, 32, 38, 58, 59, 68, 71, 85, 86, 108, 109, 111, 112, 113, 114, 115, 116, 118, 119, 121, 122, 123, 124, 127, 129, 130, 131, 150, 158, 223, 224, 225, 226, 227, 228, 229, 232, 233, 234, 235, 236, 237, 238, 242, 250, 288, 289, 290, 309
Quinton, A. 193, 194, 202

Radford, C. 75, 76, 96, 109
Ramsey, P. F. 150, 151, 152
Rantala, V. 8, 9, 16, 135, 141, 142, 146, 149, 152, 155, 158, 159, 182
Rawls, J. 11, 107, 203, 204, 206, 219
Raz, J. 266
Robinson, R. 9, 17, 108, 137, 138, 139, 140, 148, 147, 159

Rolf, B. 97
Rorty, R. 107
Russell, B. 8, 9, 17, 115, 135, 136, 148, 154, 159, 287
Ryle, G. 59, 68

Saarinen, E. 150, 159
Sam 23, 24, 25, 42
Schilpp, P. A. 131
Schlesinger, G. v, ix, 12, 13, 16, 201, 239, 283
Schwayder, D. 131
Searle, J. 9, 17, 138, 159
Sen, A. 107
Sevenonius, L. 173
Shakespeare, W. 304, 306
Shatz, D. ix, 10, 16, 187
Shelah, S. 143, 158
Shoemaker, S. 85, 86
Sibelius, P. 183
Simon, H. 163, 164, 182, 219
Skinner, B. F. 59, 68
Soames, S. 24, 50
Sober, E. 192, 193, 194, 195, 196, 202
Socrates 3, 287
Sorensen, R. 293, 294, 295, 309
Sorenson, R. 7, 15
Sosa, E. 202
Spinoza, B. 26, 106
Stalnaker, R. 83
Stavi, J. 143, 158
Stevenson, C. L. 266
Stich, S. 11, 203, 207, 208, 209, 211, 213, 214, 215, 216, 219
Strawson, P. 112, 129
Stroud, B. 202
Stump, E. 276
Summers, R. 261, 266, 267
Suppe, F. 145, 159, 202
Suppes, P. 141, 145, 158, 159, 164, 182
Svenonius, L. 144, 159, 183
Swain, M. 199, 200, 202
Swinburne, R. 275, 276, 283

Tarski, A. 9, 24, 99, 142, 155, 159, 182
Thomas, St. T. 283
Thomason, R. 83
Thorpe, D. 109
Tintoretto 288
Tomberlin, J. 202
Trollope, A. 297
Tselishchev, V. 155, 159
Tuomela, R., 144, 145, 146, 151, 154, 155, 156, 157, 158, 159, 183

Ullian, J. 309
Unger, P. 7, 71, 97, 109

Vaihinger, H. 97
van Fraassen, B. 7, 109
von Wright, G. H. 109

Wagner, S. 202
Wainwright, W. 283
Warner, R. 202
Weingartner, P. 158
Weis, J. 266
Wettstein, H. 41, 50
Wheeler, S. 97
Whitehead, A. N. 9, 17, 135, 136, 148, 154, 159
Wiener, P. 108
Wiggins, D. 85, 86
Wikler, D. 88, 108
Wittgenstein, L. 78, 91, 92, 109, 127, 292, 308
Wolterstorff, N. 276

INDEX OF SUBJECTS

a broadly behavioristic view of language 288
a complete and accurate understanding 207
A Theory of Justice 206
abbreviations 126
abbreviatory definitions 23
abortion 102
abstract objects 24, 25
actual meaning 82
actual objects 140
an absolute creator 15
analytic epistemology 203, 207, 211
analytic/synthetic distinction 111, 112, 113
analyticity 111, 112, 121, 122, 127
anti-realism 27
antireliabilists 189
appearance/stereotype/reference-fixer 291
appropriate beliefs 66
appropriate prompting conditions 224, 226
arbitrariness 214
argument form 1 256, 260
argument form 2 260
Aristotle's principle 72
Aristotle's theory of definition 3, 4
attitudinal dispositions 300
Attributability Constraint (AC) 193, 194, 195, 199

basic desiderata 71
behavioristic 59
belief 75
beliefs 11, 67
bizarre cases 87
Block Universe 25
borderline cases 72, 73, 77, 80, 81, 82, 84, 86, 88, 89, 91, 94, 95, 99, 100, 101, 106
borderline propositions 99
borderline zones 104, 105
brain states 56

Captain Horatio Hornblower 92
causal conditionals 67
causal relations 240
causal theories of reference 270
circumstantial cues 288
classical conditioning 56
clear cases 74, 75
clear zones 104, 105
cognition 55
cognition-independent modal facts 130
cognitive science 22
coherence 191
commands 126
communication 60
communication promotes cooperation 60
communities 60, 61
competence 196, 235
compositionality of vagueness 98
concept formation 4, 17
concept of justice 204
concept of justification 214
conceptions of God 270, 271, 272, 273
conceptions of justice 204
concepts 6, 52
conceptual coherentism 11, 206, 207, 208, 215, 216
conceptual necessity 116
conceptual possibilities 130
conceptual priority 12
conceptual thesis 199
concordance 103, 104

condition of adequacy 198
conditional definability 145
conditional definitions 144, 145
conditionals 67
conditioned habits 288
conditioning 56, 290
conditions of adequacy 17
conflicting opinions 90
confusion 225
conscious awareness 305
consciousness 55
conservatism 99, 101
conservative explications 208
conservative extensions 150
considered judgments 204, 205
consistency 191
context 56
contexts 7, 14, 57, 67
contingency 6
conventionalist definitions 256
conventions 7, 8, 52, 60, 61, 117, 128, 129, 130
cooperation 60
coordination 79
correcting a speaker's actual behavior 234, 235
corrigible confusion 226
counterfactuals 298, 299
Creation Theology 14, 277, 278, 279, 280, 281, 282
"creeps" 73
cues 288, 289, 290, 291, 292, 293, 294, 301

dead languages 27
"death" 88
defeasibility 261
definability 9, 10, 34, 135, 161, 162, 163, 164
definability in a model 141, 144, 173
definability in a theory 141
definability theory 147
definable in **L** 147
definiendum 122, 136, 239
definiens 122, 136, 239
defining 21, 34
defining a language 27
defining the divine 269
definition 16, 136
"definition" 112, 135
definition by example 263
definition by stipulation 121, 122, 125, 129
definition in sentences 263
definition of terms 3
definitional circularity 6, 51
definitional desiderata 97, 100
definitional methodology 16
definitional postulates 119
definitional recommendations 16
definitional regress 6, 51
definitional truths 127
definitions 45, 78, 140, 239, 287
definitions and definability ix
definitions in law 253, 265
definitions in mathematics 137, 138
definitions in the law 13
definitions in theology 16
denotations 22, 24
deontic logic 13, 245, 246, 247, 248, 249
describing usage 89
descriptions 12
desiderata for definitions 71
determinate meaning 8
dialectical skepticism 10, 16, 190, 192
dialectically effective anti-skeptical arguments 195
dicionaries 117
dictionaries 35
dictionary definitions 51
different definitions 36
differentia 3, 148
disjunctive physical traits 229
dispositional account of logic 223
dispositional conception of language 16
dispositional conception of logic 12
dispositional differences 58, 59
dispositional predicates 145
dispositions 7, 17, 52, 59, 66, 224, 226, 227, 228, 230, 236, 298, 299, 308
dispositions to assent 294
dispositions to dissent 294
dispositions toward behavior 56

dissent 224
distinct properties 36
"Don't ask for the meaning, ask for the use" 21
"dudent" 120, 123

economy 54
elegance 54
eliminability 9, 148, 154
eliminability in a model **M** 152
elminativism 293
embeddability as a test for vagueness 73
emotivism 255
empirical discovery 200
empirical inquiry 16
empirical results 161
empirical science 17
empirical analysis 4, 5
energy 27
"entails" 228
entities 42
environmental factors 61
epistemic interdependence 66
epistemic justification 207
epistemic justifications 11
epistemic positions 191
epistemic rationality 211, 212, 213, 214, 216, 217
epistemic skepticism 10, 190, 196
epistemic terms 16
epistemic theorists 102, 103
epistemic xenophobia 209, 210, 211
Epistemology and Cognition 200
equivalence 140
equivalent definitions 36
essential properties 3, 5
essentially contested concepts 92, 109
event identity 12, 13, 240
events 12, 13, 240
exemplifications of properties at times 12
explication 5, 131
explications 81
explicative definitions 81
explicit definability 10, 142, 172
explicit definitions 15
explicit identifiability 10, 175
expressing 113
extended interpolation theorem 174
extensional equivalence 298
extensionalism 59
external world 66

factual claims 27
Faith and Knowledge 275
false determinacies 87
fetus 14
finite definability 143
finite identifiability 143
first languages 288
"follow deductively from" 227
"for the sake of convenience" 43
formal languages 119
formal methods 9, 135
formal models 16
formal systems 37
formalized language 32
fundamental unit of meaning 7
fuzziness 78

game-theoretic rules for interrogative derivability 10
games 41
garlic 62, 64
"gavagai" 58, 308
general laws 28, 30, 32
generality 73, 77
generic rationality 211, 212, 213, 214, 215, 216, 217
genetic factors 61
genuine synonymy 119
genus 3, 148
global definability 9, 146
global uncertainty 146
God as all-knowing 271
God as all-powerful 271
God as creator of the world 271
God as eternal 271
God as not like anything in the world 272
God as perfectly good 271
God as transcendent 271
God as unlike anything in the world 272
God not in space-time 271

"going on the same way" 292
greatest possible being 15

habits 7, 52, 66
Hamlet 304
Happiness 218
hard cases 74, 76, 287
Heisenberg's uncertainty principle 106
hidden truth values 84
human communities 60
hypothetical robot, Karl 289

icons 55, 56, 63
idealized behavior 234
idealized competence 235
idealized concepts in physics 229, 230, 231, 232, 233, 237
idealized conditions 234
idealized dispositions 16, 223, 229, 232
idealized laws 231
idealized notions 11
idealized notions in physics 12
idealized speech dispositions 12
idealized toward a norm 12, 235, 236
identical events 240, 241
identifiability 9, 10, 142, 161, 162, 163, 164, 165
identification 16
identity 140
idiolects 289
ignorance about appearances 291
Iliad 305
illocutionary force 138
implication 33
implicit definability 142, 172
implicit identifiability 10
implicity definability 10
incoherence theorists 81
incoherentism 95, 97, 98, 99, 109
incoherentists 77, 96
incompleteness 84
inconsistency 71, 95, 99
indeterminacies of the definiendum 78
indeterminacies of the definiens 78
indeterminacy of meaning 79, 86
indeterminacy of translation 8, 111, 113
indeterminate cases 92
indices 55, 56, 63
indirect quotation 43
inductive regularities 236, 237
"infelicities" 139
inferable 35
inference to the best explanation 195
information 217, 218
innateness 308
instrumental 257
instrumental definitions 263, 265
instrumentalism 256
"intelligent" 290, 291
intentions of definitions 9
interchangeability 24
interpolation theorem 171
interrogative derivability 169
interrogative inquiry 180, 181, 182
interrogative model of inquiry 10, 165
interrogatory model 16
intrinsic similarity 292
introducing a term 22
introspective reports 292, 293
intuitions 10, 11, 187
invalid inference 238
inverted definitions 12, 16
irresolvable disagreements 92

judicial function 16
justice 11
"justice" 204
justifiabile legal system 14
justification 207
"justified" 198
justified true belief 187

"kinetic energy" 27, 28, 29, 30
KK principle 75
"know" 296
knowledge 10, 11
"knowledge" 75
"knowledge implies belief" 96

lacking an opinion 90
language 6, 31, 35
language a system of conditioned habits 288
language learning 308

language of thought 53, 54, 56
languages as abstract objects 24
law of excluded middle 120
law of non-contradiction 120
lawlike sentences 67
laws 28, 30, 32, 52, 64, 298
laws of nature 28, 30, 32, 52, 64, 78, 298
learning a language 15
legal consequences of definitions 14
legal instrumentalism 256
legal norms vs. ethical norms 265
legal realism 255, 256
Lewis' theory of conventions 128
"life" 88
linguistic analysis 208
linguistic communities 85, 91
linguistic data 90
linguistic primitives 53
linguistic understanding 54
local definability 9, 146
local meaninglessness 83, 84
local uncertainty 146
Logical Atomism 49
logical constants 120
Logical Empiricism 145
logical implications 170
logical inference 35
Logical Positivism 8, 91, 116, 117, 119
logical truth 118, 123
logical truths 12, 38, 223, 224, 225, 226, 233

Macbeth 303
many-valued logics 102
mass in classical mechanics 9, 10
mathematical definitions 137, 138
maximization of truth 16
maximizing rationality 212
maximizing true beliefs 11, 206
meaning 22, 31, 67
meaning analysis 4
meaning of a sign 56
meaning of a symbol 56
meaning of "gavagai" 58
meaning of "meaning" 33
meaning of primitives 60
meaning postulates 119, 124, 131
meaning preservation 100, 107
meaning rules 33, 34, 35
meaning without analyticity 120
meaningful brain states 56
meaningful but vague predicates 72
meaninglessness 83, 84
meanings 21
meanings guarantee truth 8
measurability 10
measureability 179
Mentalese 53, 54, 56, 288
mentality 16
metaphor 21, 72
methodology in philosophy ix, 3
mind-dependence of language 15
mind/body problem 106
minds 55, 63
minds of Type I 63
minds of Type II 63
minds of Type III 63
minimization of falsity 16
minimizing false beliefs 11, 206
miracles 78
model theory 140
model-theoretic distinctions 140
momentum 20
monotheism 14, 274
moral obligations 243, 244
murder 255

n-foldly identifiable 143
natural kind terms 257, 265
natural language 16
natural laws 6, 7, 52, 60, 61, 64, 298
natural signs 63
naturalistic accounts 223
naturalistic conception of logic 12
naturalistic philosophy 11, 235
"Nature" 165
nature of meaning 3
necessary truth 37
necessary truths 26, 29, 30
necessity 6
nervous systems 288, 289
nomic properties 5

nominal definitions 4, 6, 9, 21, 23, 25, 30, 31, 32, 34, 39, 136, 137, 140, 156, 157
nominalism 46
non-conservative extensions 37
non-linguistic understanding 54
non-synonymies 80
noncreativity 9, 148, 150, 154
nonlogical axioms 131
norms 12

objects 13
"observable" 178
observability 10, 179
"observability" 178
"obvious implication" 228
Odyssey 305
omnipotence 277
omnipotent creator 14
omniscience 277
omniscient creator 14
ontology 42
operant conditioning 56
operational definitions 107
opportunities 55
optimistic incoherentists 97
ordinary beliefs 10, 11
ordinary believer constraint #1 194
ordinary believer constraint #2 196, 197, 198
ordinary language philosophers 72
Othello 304

pantheism 14, 274
"paraphrase" 114
Paris, A Rainy Day 303
parity thesis 80
partisan borderline cases 90, 94
pejorative meaning 31
perception 291, 292, 296P
Perfect Being Theology 14, 277, 278, 279, 281, 282
performance 196
performative conditions 9
performative utterances 139
performatory 22
personal habits 6, 52, 60, 61
personal survival 84, 85, 89
pessimistic incoherentists 96
philosophical analyses 15, 16, 287, 295, 308
philosophical definitions 287, 297
philosophical methodology ix
philosophical skepticism 188
philosophy 17
Philosophy of Logic 224
piecewise definability 10, 144, 153, 172
piecewise eliminability 153
piecewise explicity identifiability 175
piecewise identifiability 10
Plato's theory of definition 79
Platonism 25, 27, 41, 46, 47, 293, 309
polytheism 14, 274
possible worlds 8, 111, 140
possible worlds semantics 16
potential meaning 82
practical rationality 219
pragmatic aspects 135, 138
pragmatic inference 35
pragmatical elements 16
pre-theoretical notions 204
precisification 72, 74, 81, 104
precisifications 88, 89, 102
precisifying definitions 8
precision 71, 77
precision in definition 16
prescribing usage 89
prescriptivism 255
presuppositions 169
primitive concepts 51, 52, 54
primitive words 6, 52
Principia Mathematica 135, 136, 148, 149, 154, 156
principle (D) 239, 240, 241
private languages 59
private meaning 65
probability 77, 108
prompting conditions 12, 224, 225, 226
proof procedures 226
proper definitions 3
proper prompting conditions 225, 226
proper usage 12
properties 21, 22, 23, 24, 25, 27, 33, 35, 36, 40, 41, 42, 48

propositions 21, 22, 23, 24, 25, 28, 29, 40, 43, 44, 114
proximal stimuli 289
psychological primitives 53
psychology 15
public usage 65
pure cases 231
purely nominal definitions 23, 24
purposes of definitions 13, 137, 253, 263, 264, 265

Quine's thesis *T* 223, 232, 237
Quinean conception of language 17

Ramsey-eliminability 150, 151
random events 13
range of indeterminacy 294
rational beliefs 192
rational reconstruction 100
rationality 211
real definitions 4, 5, 6, 9, 21, 23, 25, 38, 48, 137, 138, 140, 156
realism 13, 255, 256, 259
realist attitudes toward definitions 13
realist definitions 257, 258
reality/reference/rigid-designator 291
reducibility 149
reductionism 50
reference without meaning 15
reflective equilibrium 11, 16, 203, 208, 209, 213
Regina v. Objibway 253, 258
regressive definitions 51
reliabilism 188, 189, 190, 192, 197, 201, 202
reliabilist view of knowledge 10
reliabilists 189
restricted definability 144
restricted eliminability 152
restricted identifiability 144
righteous individuals 13, 243, 244, 250
rules 126
rules for introducing a term 31
rules of meaning 32, 33, 34, 35
rules of proper usage 236

satisficing 219, 220
satisficing approach 11, 16
satisficing rationality 212
scientific concepts 82
semantic discretionism 86, 87, 88, 89, 99
semantic incompleteness 84
semantic pragmatisim 85
semantical rules 119
semantics 15, 22
semiotic 55
semiotic significance for a system at a time 67
semiotic systems 7, 55, 67
sensations 93
sets of possible worlds 8, 111
shared dispositions 52
shared habits 7, 52
shared tendencies 52
Sherwood v. Walker 254
sign-using systems 7, 55
similarity 292
similarity of meaning 56
simplicity 54
simplification 106
skepticism 10, 193
skepticism as dialectical 11
smell of garlic 62
smells 64
social conventions 6, 7, 52, 60, 61, 64
solvable cases 74
sorites 47
sorites embeddability 73
sorties 73
speaker's actual behavior 235
speaker's competence 236
speaker's dispositions 236
speakers' assents 12
speakers' dissents 12
species 3
speech act theory 135
split-brain research 95
spontaneous events 13
stimulus meaning 289
stipulation 100
stipulations 21, 28, 29, 126
stipulative definitions 8, 24, 27, 31, 131, 138, 139, 157

stipulative nominal definitions 26
stipulatives 125, 127
strong implicit identifiability 175
subjunctive conditionals 67
subtableau 167, 168, 169, 171
supervaluationism 83
supervaluationist position 8
supervaluationists 77, 82, 105
symbols 55, 56, 63, 135, 137
synonymies 80
synonymy 9, 121
"synonymy" 112
syntactic form 140
Syntactic Structures 89
syntactical forms of definitions 9
systematic power 54

tableau 165, 166, 169
Tarski's theory of truth 24
tendencies 7, 52, 66
tendencies toward behavior 7
testability 71
testing hypotheses about meaning 66
tests 78
The Coherence of Theism 275
the concept of God 269
the concept of justice 204
the discovery argument 10, 11, 201
The Eustace Diamonds 297
the existence of an external world 65
the goal of criticism 303
"the Inquirer" 165, 183
the language of thought 53, 54, 56
the nature of belief 75
the nature of knowledge 75
the Null proposition 114, 130
the ordinary believer argument 10, 192, 201
the possibility argument 10, 188, 189, 200
The Rape of the Lock 306
the simple account 113, 114, 115
the Verificationist account 115, 116
theory construction 4, 17, 71
theory of definability 155
theory of definition 3
theory of derivability 155
theory of knowleldge 187
theory of mind 7
theory of signs 7, 55
theory of truth 24
things 137
thought experiments 15, 71, 300, 301
"to a certain extent" 43
traditional conceptions of God 14
traditions in theology 274, 275, 276, 277
translation procedure 232, 233
triadic conception of language 16
triggers of verbal dispositions 298
true belief 206
true by definition 124
true by virtue of meaning 30
truth 16, 28, 66, 78, 99, 120
truth as appropriate belief 66
truth by convention 8, 117, 120, 127, 129, 130
truth by experiment 28
truth by stipulation 8, 28, 117
truth by virtue of meaning 111
truth condition semantics 53, 54
truth conditions 24, 53, 81
truth of C in M 170
truth value 28
truth values 84
truth-conditional semantics 108
twin water 291
two aspects of use 31
"Two Dogmas of Empiricism" 111, 112, 113, 122
type I minds 63
type II minds 63
type III minds 63
type/token recognition 56
types 25

unacceptable cognitive states 11
uncertainty 180
undefeated justified true belief 188
underspecificity 76, 77
understanding 12, 13, 242
unfalsifiability 8, 115
uniformity of usage 83

universal laws 230, 231
unrevisability 112
utilitarianism 43, 44
utterances 288

vague expressions 7
vagueness 7, 15, 35, 71, 72, 73, 76, 77, 83, 84, 89, 90, 91, 108
vaguification 104, 105
valid inference 238
"veline" 125, 126, 127
verbal behavior 12, 15
verbal cues 15
verbal dispositions 298, 300
verification conditions 116

water 291
weak implicit identifiability 176
Word and Object 111

zenophobia 210

Zettel 92

SYNTHESE LIBRARY

Studies in Epistemology, Logic, Methodology, and Philosophy of Science

1. J. M. Bochénski, *A Precis of Mathematical Logic*. 1959 ISBN 90-277-0073-7
2. P. Guiraud, *Problèmes et méthodes de la statistique linguistique*. 1959 ISBN 90-277-0025-7
3. H. Freudenthal (ed.), *The Concept and the Role of the Model in Mathematics and Natural and Social Sciences*. 1961 ISBN 90-277-0017-6
4. E. W. Beth, *Formal Methods*. An Introduction to Symbolic Logic and to the Study of Effective Operations in Arithmetic and Logic. 1962 ISBN 90-277-0069-9
5. B. H. Kazemier and D. Vuysje (eds.), *Logic and Language*. Studies dedicated to Professor Rudolf Carnap on the Occasion of His 70th Birthday. 1962 ISBN 90-277-0019-2
6. M. W. Wartofsky (ed.), *Proceedings of the Boston Colloquium for the Philosophy of Science, 1961–1962*. [Boston Studies in the Philosophy of Science, Vol. I] 1963 ISBN 90-277-0021-4
7. A. A. Zinov'ev, *Philosophical Problems of Many-valued Logic*. A revised edition, edited and translated (from Russian) by G. Küng and D.D. Comey. 1963 ISBN 90-277-0091-5
8. G. Gurvitch, *The Spectrum of Social Time*. Translated from French and edited by M. Korenbaum and P. Bosserman. 1964 ISBN 90-277-0006-0
9. P. Lorenzen, *Formal Logic*. Translated from German by F.J. Crosson. 1965 ISBN 90-277-0080-X
10. R. S. Cohen and M. W. Wartofsky (eds.), *Proceedings of the Boston Colloquium for the Philosophy of Science, 1962–1964*. In Honor of Philipp Frank. [Boston Studies in the Philosophy of Science, Vol. II] 1965 ISBN 90-277-9004-0
11. E. W. Beth, *Mathematical Thought*. An Introduction to the Philosophy of Mathematics. 1965 ISBN 90-277-0070-2
12. E. W. Beth and J. Piaget, *Mathematical Epistemology and Psychology*. 1966 ISBN 90-277-0071-0
13. G. Küng, *Ontology and the Logistic Analysis of Language*. An Enquiry into the Contemporary Views on Universals. 1967 ISBN 90-277-0028-1
14. R. S. Cohen and M. W. Wartofsky (eds.), *Proceedings of the Boston Colloquium for the Philosophy of Sciences, 1964–1966*. In Memory of Norwood Russell Hanson. [Boston Studies in the Philosophy of Science, Vol. III] 1967 ISBN 90-277-0013-3
15. C. D. Broad, *Induction, Probability, and Causation*. Selected Papers. 1968 ISBN 90-277-0012-5
16. G. Patzig, *Aristotle's Theory of the Syllogism*. A Logical-philosophical Study of *Book A* of the *Prior Analytics*. 1968 ISBN 90-277-0030-3

17. N. Rescher, *Topics in Philosophical Logic*. 1968 ISBN 90-277-0084-2
18. R. S. Cohen and M. W. Wartofsky (eds.), *Proceedings of the Boston Colloquium for the Philosophy of Science, 1966–1968, Part I*. [Boston Studies in the Philosophy of Science, Vol. IV] 1969 ISBN 90-277-0014-1
19. R. S. Cohen and M. W. Wartofsky (eds.), *Proceedings of the Boston Colloquium for the Philosophy of Science, 1966–1968, Part II*. [Boston Studies in the Philosophy of Science, Vol. V] 1969 ISBN 90-277-0015-X
20. J. W. Davis, D. J. Hockney and W. K. Wilson (eds.), *Philosophical Logic*. 1969 ISBN 90-277-0075-3
21. D. Davidson and J. Hintikka (eds.), *Words and Objections*. Essays on the Work of W. V. Quine. 1969 ISBN 90-277-0074-5; Pb 90-277-0602-6
22. P. Suppes, *Studies in the Methodology and Foundations of Science. Selected Papers from 1911 to 1969*. 1969 ISBN 90-277-0020-6
23. J. Hintikka, *Models for Modalities*. Selected Essays. 1969 ISBN 90-277-0078-8; Pb 90-277-0598-4
24. N. Rescher *et al.* (eds.), *Essays in Honor of Carl G. Hempel*. A Tribute on the Occasion of His 65th Birthday. 1969 ISBN 90-277-0085-0
25. P. V. Tavanec (ed.), *Problems of the Logic of Scientific Knowledge*. 1969 ISBN 90-277-0087-7
26. M. Swain (ed.), *Induction, Acceptance, and Rational Belief*. 1970 ISBN 90-277-0086-9
27. R. S. Cohen and R. J. Seeger (eds.), *Ernst Mach: Physicist and Philosopher*. [Boston Studies in the Philosophy of Science, Vol. VI]. 1970 ISBN 90-277-0016-8
28. J. Hintikka and P. Suppes, *Information and Inference*. 1970 ISBN 90-277-0155-5
29. K. Lambert, *Philosophical Problems in Logic*. Some Recent Developments. 1970 ISBN 90-277-0079-6
30. R. A. Eberle, *Nominalistic Systems*. 1970 ISBN 90-277-0161-X
31. P. Weingartner and G. Zecha (eds.), *Induction, Physics, and Ethics*. 1970 ISBN 90-277-0158-X
32. E. W. Beth, *Aspects of Modern Logic*. 1970 ISBN 90-277-0173-3
33. R. Hilpinen (ed.), *Deontic Logic*. Introductory and Systematic Readings. 1971 ISBN 90-277-0167-9; Pb 90-277-1302-2
34. J.-L. Krivine, *Introduction to Axiomatic Set Theory*. 1971 ISBN 90-277-0169-5; Pb 90-277-0411-2
35. J. D. Sneed, *The Logical Structure of Mathematical Physics*. 2nd rev. ed., 1979 ISBN 90-277-1056-2; Pb 90-277-1059-7
36. C. R. Kordig, *The Justification of Scientific Change*. 1971 ISBN 90-277-0181-4; Pb 90-277-0475-9
37. M. Čapek, *Bergson and Modern Physics*. A Reinterpretation and Re-evaluation. [Boston Studies in the Philosophy of Science, Vol. VII] 1971 ISBN 90-277-0186-5
38. N. R. Hanson, *What I Do Not Believe, and Other Essays*. Ed. by S. Toulmin and H. Woolf. 1971 ISBN 90-277-0191-1
39. R. C. Buck and R. S. Cohen (eds.), *PSA 1970*. Proceedings of the Second Biennial Meeting of the Philosophy of Science Association, Boston, Fall 1970. In Memory of Rudolf Carnap. [Boston Studies in the Philosophy of Science, Vol. VIII] 1971 ISBN 90-277-0187-3; Pb 90-277-0309-4
40. D. Davidson and G. Harman (eds.), *Semantics of Natural Language*. 1972 ISBN 90-277-0304-3; Pb 90-277-0310-8

41. Y. Bar-Hillel (ed.), *Pragmatics of Natural Languages*. 1971
ISBN 90-277-0194-6; Pb 90-277-0599-2
42. S. Stenlund, *Combinators, λ-Terms and Proof Theory*. 1972 ISBN 90-277-0305-1
43. M. Strauss, *Modern Physics and Its Philosophy*. Selected Paper in the Logic, History, and Philosophy of Science. 1972 ISBN 90-277-0230-6
44. M. Bunge, *Method, Model and Matter*. 1973 ISBN 90-277-0252-7
45. M. Bunge, *Philosophy of Physics*. 1973 ISBN 90-277-0253-5
46. A. A. Zinov'ev, *Foundations of the Logical Theory of Scientific Knowledge (Complex Logic)*. Revised and enlarged English edition with an appendix by G. A. Smirnov, E. A. Sidorenka, A. M. Fedina and L. A. Bobrova. [Boston Studies in the Philosophy of Science, Vol. IX] 1973 ISBN 90-277-0193-8; Pb 90-277-0324-8
47. L. Tondl, *Scientific Procedures*. [Boston Studies in the Philosophy of Science, Vol. X] 1973 ISBN 90-277-0147-4; Pb 90-277-0323-X
48. N. R. Hanson, *Constellations and Conjectures*. Ed. by W. C. Humphreys, Jr. 1973
ISBN 90-277-0192-X
49. K. J. J. Hintikka, J. M. E. Moravcsik and P. Suppes (eds.), *Approaches to Natural Language*. 1973 ISBN 90-277-0220-9; Pb 90-277-0233-0
50. M. Bunge (ed.), *Exact Philosophy*. Problems, Tools and Goals. 1973
ISBN 90-277-0251-9
51. R. J. Bogdan and I. Niiniluoto (eds.), *Logic, Language and Probability*. 1973
ISBN 90-277-0312-4
52. G. Pearce and P. Maynard (eds.), *Conceptual Change*. 1973
ISBN 90-277-0287-X; Pb 90-277-0339-6
53. I. Niiniluoto and R. Tuomela, *Theoretical Concepts and Hypothetico-inductive Inference*. 1973 ISBN 90-277-0343-4
54. R. Fraissé, *Course of Mathematical Logic* – Volume 1: *Relation and Logical Formula*. 1973 ISBN 90-277-0268-3; Pb 90-277-0403-1
(For *Volume 2* see under Nr. 69).
55. A. Grünbaum, *Philosophical Problems of Space and Time*. 2nd enlarged ed. [Boston Studies in the Philosophy of Science, Vol. XII] 1973
ISBN 90-277-0357-4; Pb 90-277-0358-2
56. P. Suppes (ed.), *Space, Time and Geometry*. 1973
ISBN 90-277-0386-8; Pb 90-277-0442-2
57. H. Kelsen, *Essays in Legal and Moral Philosophy*. Selected and introduced by O. Weinberger. 1973 ISBN 90-277-0388-4
58. R. J. Seeger and R. S. Cohen (eds.), *Philosophical Foundations of Science*. [Boston Studies in the Philosophy of Science, Vol. XI] 1974
ISBN 90-277-0390-6; Pb 90-277-0376-0
59. R. S. Cohen and M. W. Wartofsky (eds.), *Logical and Epistemological Studies in Contemporary Physics*. [Boston Studies in the Philosophy of Science, Vol. XIII] 1973 ISBN 90-277-0391-4; Pb 90-277-0377-9
60. R. S. Cohen and M. W. Wartofsky (eds.), *Methodological and Historical Essays in the Natural and Social Sciences. Proceedings of the Boston Colloquium for the Philosophy of Science, 1969–1972*. [Boston Studies in the Philosophy of Science, Vol. XIV] 1974 ISBN 90-277-0392-2; Pb 90-277-0378-7
61. R. S. Cohen, J. J. Stachel and M. W. Wartofsky (eds.), *For Dirk Struik. Scientific, Historical and Political Essays*. [Boston Studies in the Philosophy of Science, Vol. XV] 1974 ISBN 90-277-0393-0; Pb 90-277-0379-5

62. K. Ajdukiewicz, *Pragmatic Logic*. Transl. from Polish by O. Wojtasiewicz. 1974
ISBN 90-277-0326-4
63. S. Stenlund (ed.), *Logical Theory and Semantic Analysis*. Essays dedicated to Stig Kanger on His 50th Birthday. 1974 ISBN 90-277-0438-4
64. K. F. Schaffner and R. S. Cohen (eds.), *PSA 1972. Proceedings of the Third Biennial Meeting of the Philosophy of Science Association*. [Boston Studies in the Philosophy of Science, Vol. XX] 1974 ISBN 90-277-0408-2; Pb 90-277-0409-0
65. H. E. Kyburg, Jr., *The Logical Foundations of Statistical Inference*. 1974
ISBN 90-277-0330-2; Pb 90-277-0430-9
66. M. Grene, *The Understanding of Nature*. Essays in the Philosophy of Biology. [Boston Studies in the Philosophy of Science, Vol. XXIII] 1974
ISBN 90-277-0462-7; Pb 90-277-0463-5
67. J. M. Broekman, *Structuralism: Moscow, Prague, Paris*. 1974
ISBN 90-277-0478-3
68. N. Geschwind, *Selected Papers on Language and the Brain*. [Boston Studies in the Philosophy of Science, Vol. XVI] 1974 ISBN 90-277-0262-4; Pb 90-277-0263-2
69. R. Fraissé, *Course of Mathematical Logic* – Volume 2: *Model Theory*. 1974
ISBN 90-277-0269-1; Pb 90-277-0510-0
(For *Volume 1* see under Nr. 54)
70. A. Grzegorczyk, *An Outline of Mathematical Logic*. Fundamental Results and Notions Explained with All Details. 1974 ISBN 90-277-0359-0
71. F. von Kutschera, *Philosophy of Language*. 1975 ISBN 90-277-0591-7
72. J. Manninen and R. Tuomela (eds.), *Essays on Explanation and Understanding*. Studies in the Foundations of Humanities and Social Sciences. 1976
ISBN 90-277-0592-5
73. J. Hintikka (ed.), *Rudolf Carnap, Logical Empiricist*. Materials and Perspectives. 1975 ISBN 90-277-0583-6
74. M. Čapek (ed.), *The Concepts of Space and Time*. Their Structure and Their Development. [Boston Studies in the Philosophy of Science, Vol. XXII] 1976
ISBN 90-277-0355-8; Pb 90-277-0375-2
75. J. Hintikka and U. Remes, *The Method of Analysis*. Its Geometrical Origin and Its General Significance. [Boston Studies in the Philosophy of Science, Vol. XXV] 1974 ISBN 90-277-0532-1; Pb 90-277-0543-7
76. J. E. Murdoch and E. D. Sylla (eds.), *The Cultural Context of Medieval Learning*. [Boston Studies in the Philosophy of Science, Vol. XXVI] 1975
ISBN 90-277-0560-7; Pb 90-277-0587-9
77. S. Amsterdamski, *Between Experience and Metaphysics*. Philosophical Problems of the Evolution of Science. [Boston Studies in the Philosophy of Science, Vol. XXXV] 1975 ISBN 90-277-0568-2; Pb 90-277-0580-1
78. P. Suppes (ed.), *Logic and Probability in Quantum Mechanics*. 1976
ISBN 90-277-0570-4
79. H. von Helmholtz: *Epistemological Writings. The Paul Hertz/Moritz Schlick Centenary Edition of 1921 with Notes and Commentary by the Editors*. Newly translated by M. F. Lowe. Edited, with an Introduction and Bibliography, by R. S. Cohen and Y. Elkana. [Boston Studies in the Philosophy of Science, Vol. XXXVII] 1975 ISBN 90-277-0290-X; Pb 90-277-0582-8
80. J. Agassi, *Science in Flux*. [Boston Studies in the Philosophy of Science, Vol. XXVIII] 1975 ISBN 90-277-0584-4; Pb 90-277-0612-2

81. S. G. Harding (ed.), *Can Theories Be Refuted?* Essays on the Duhem-Quine Thesis. 1976 ISBN 90-277-0629-8; Pb 90-277-0630-1
82. S. Nowak, *Methodology of Sociological Research.* General Problems. 1977 ISBN 90-277-0486-4
83. J. Piaget, J.-B. Grize, A. Szemińska, and V. Bang, *Epistemology and Psychology of Functions.* 1977 ISBN 90-277-0804-5
84. M. Grene and E. Mendelsohn (eds.), *Topics in the Philosophy of Biology.* [Boston Studies in the Philosophy of Science, Vol. XXVII] 1976 ISBN 90-277-0595-X; Pb 90-277-0596-8
85. E. Fischbein, *The Intuitive Sources of Probabilistic Thinking in Children.* 1975 ISBN 90-277-0626-3
86. E. W. Adams, *The Logic of Conditionals.* An Application of Probability to Deductive Logic. 1975 ISBN 90-277-0631-X
87. M. Przełęcki and R. Wójcicki (eds.), *Twenty-Five Years of Logical Methodology in Poland.* 1976 ISBN 90-277-0601-8
88. J. Topolski, *The Methodology of History.* 1976 ISBN 90-277-0550-X
89. A. Kasher (ed.), *Language in Focus: Foundations, Methods and Systems.* Essays dedicated to Yehoshua Bar-Hillel. [Boston Studies in the Philosophy of Science, Vol. XLIII] 1976 ISBN 90-277-0644-1; Pb 90-277-0645-X
90. J. Hintikka, *The Intentions of Intentionality and Other New Models for Modalities.* 1975 ISBN 90-277-0633-6; Pb 90-277-0634-4
91. W. Stegmüller, *Collected Papers on Epistemology, Philosophy of Science and History of Philosophy.* 2 Volumes. 1977 Set ISBN 90-277-0767-7
92. D. M. Gabbay, *Investigations in Modal and Tense Logics with Applications to Problems in Philosophy and Linguistics.* 1976 ISBN 90-277-0656-5
93. R. J. Bogdan, *Local Induction.* 1976 ISBN 90-277-0649-2
94. S. Nowak, *Understanding and Prediction.* Essays in the Methodology of Social and Behavioral Theories. 1976 ISBN 90-277-0558-5
95. P. Mittelstaedt, *Philosophical Problems of Modern Physics.* [Boston Studies in the Philosophy of Science, Vol. XVIII] 1976 ISBN 90-277-0285-3; Pb 90-277-0506-2
96. G. Holton and W. A. Blanpied (eds.), *Science and Its Public: The Changing Relationship.* [Boston Studies in the Philosophy of Science, Vol. XXXIII] 1976 ISBN 90-277-0657-3; Pb 90-277-0658-1
97. M. Brand and D. Walton (eds.), *Action Theory.* 1976 ISBN 90-277-0671-9
98. P. Gochet, *Outline of a Nominalist Theory of Proposition.* An Essay in the Theory of Meaning. 1980 ISBN 90-277-1031-7
99. R. S. Cohen, P. K. Feyerabend, and M. W. Wartofsky (eds.), *Essays in Memory of Imre Lakatos.* [Boston Studies in the Philosophy of Science, Vol. XXXIX] 1976 ISBN 90-277-0654-9; Pb 90-277-0655-7
100. R. S. Cohen and J. J. Stachel (eds.), *Selected Papers of Léon Rosenfield.* [Boston Studies in the Philosophy of Science, Vol. XXI] 1979 ISBN 90-277-0651-4; Pb 90-277-0652-2
101. R. S. Cohen, C. A. Hooker, A. C. Michalos and J. W. van Evra (eds.), *PSA 1974. Proceedings of the 1974 Biennial Meeting of the Philosophy of Science Association.* [Boston Studies in the Philosophy of Science, Vol. XXXII] 1976 ISBN 90-277-0647-6; Pb 90-277-0648-4
102. Y. Fried and J. Agassi, *Paranoia.* A Study in Diagnosis. [Boston Studies in the Philosophy of Science, Vol. L] 1976 ISBN 90-277-0704-9; Pb 90-277-0705-7

103. M. Przełęcki, K. Szaniawski and R. Wójcicki (eds.), *Formal Methods in the Methodology of Empirical Sciences*. 1976 ISBN 90-277-0698-0
104. J. M. Vickers, *Belief and Probability*. 1976 ISBN 90-277-0744-8
105. K. H. Wolff, *Surrender and Catch*. Experience and Inquiry Today. [Boston Studies in the Philosophy of Science, Vol. LI] 1976 ISBN 90-277-0758-8; Pb 90-277-0765-0
106. K. Kosík, *Dialectics of the Concrete*. [Boston Studies in the Philosophy of Science, Vol. LII] 1976 ISBN 90-277-0761-8; Pb 90-277-0764-2
107. N. Goodman, *The Structure of Appearance*. 3rd ed., 1977 [Boston Studies in the Philosophy of Science, Vol. LIII] 1977 ISBN 90-277-0773-1; Pb 90-277-0774-X
108. Kazimierz Ajdukiewicz, *The Scientific World-Perspective and Other Essays, 1931-1963*. Edited and with an Introduction by J. Giedymin. 1978 ISBN 90-277-0527-5
109. R. L. Causey, *Unity of Science*. 1977 ISBN 90-277-0779-0
110. R. E. Grandy, *Advanced Logic for Applications*. 1977 ISBN 90-277-0781-2
111. R. P. McArthur, *Tense Logic*. 1976 ISBN 90-277-0697-2
112. L. Lindahl, *Position and Change*. A Study in Law and Logic. 1977 ISBN 90-277-0787-1
113. R. Tuomela, *Dispositions*. 1978 ISBN 90-277-0810-X
114. H. A. Simon, *Models of Discovery and Other Topics in the Methods of Science*. [Boston Studies in the Philosophy of Science, Vol. LIV] 1977 ISBN 90-277-0812-6; Pb 90-277-0858-4
115. R. D. Rosenkrantz, *Inference, Method and Decision*. Towards a Bayesian Philosophy of Science. 1977 ISBN 90-277-0817-7; Pb 90-277-0818-5
116. R. Tuomela, *Human Action and Its Explanation*. A Study on the Philosophical Foundations of Psychology. 1977 ISBN 90-277-0824-X
117. M. Lazerowitz, *The Language of Philosophy*. Freud and Wittgenstein. [Boston Studies in the Philosophy of Science, Vol. LV]. 1977 ISBN 90-277-0826-6; Pb 90-277-0862-2
118. Not published
119. J. Pelc, *Semiotics in Poland, 1894–1969*. 1979 ISBN 90-277-0811-8
120. I. Pörn, *Action Theory and Social Science*. Some Formal Models. 1977 ISBN 90-277-0846-0
121. J. Margolis, *Persons and Mind*. The Prospects of Nonreductive Materialism. [Boston Studies in the Philosophy of Science, Vol. LVII]. 1977 ISBN 90-277-0854-1; Pb 90-277-0863-0
122. J. Hintikka, I. Niiniluoto, and E. Saarinen (eds.), *Essays on Mathematical and Philosophical Logic*. 1979 ISBN 90-277-0879-7
123. T. A. F. Kuipers, *Studies in Inductive Probability and Rational Expectation*. 1978 ISBN 90-277-0882-7
124. E. Saarinen, R. Hilpinen, I. Niiniluoto and M. P. Hintikka (eds.), *Essays in Honour of Jaakko Hintikka on the Occasion of His 50th Birthday*. 1979 ISBN 90-277-0916-5
125. G. Radnitzky and G. Andersson (eds.), *Progress and Rationality in Science*. [Boston Studies in the Philosophy of Science, Vol. LVIII] 1978 ISBN 90-277-0921-1; Pb 90-277-0922-X
126. P. Mittelstaedt, *Quantum Logic*. 1978 ISBN 90-277-0925-4
127. K. A. Bowen, *Model Theory for Modal Logic*. Kripke Models for Modal Predicate Calculi. 1979 ISBN 90-277-0929-7

128. H. A. Bursen, *Dismantling the Memory Machine.* A Philosophical Investigation of Machine Theories of Memory. 1978 ISBN 90-277-0933-5
129. M. W. Wartofsky, *Models*, Representation and the Scientific Understanding. [Boston Studies in the Philosophy of Science, Vol. XLVIII.] 1979 ISBN 90-277-0736-7; Pb 90-277-0947-5
130. D. Ihde, *Technics and Praxis.* A Philosophy of Technology. [Boston Studies in the Philosophy of Science, Vol. XXIV] 1979 ISBN 90-277-0953-X; Pb 90-277-0954-8
131. J. J. Wiatr (ed.), *Polish Essays in the Methodology of the Social Sciences.* [Boston Studies in the Philosophy of Science, Vol. XXIX] 1979 ISBN 90-277-0723-5; Pb 90-277-0956-4
132. W. C. Salmon (ed.), *Hans Reichenbach: Logical Empiricist.* 1979 ISBN 90-277-0958-0
133. P. Bieri, R.-P. Horstmann and L. Krüger (eds.), *Transcendental Arguments in Science.* Essays in Epistemology. 1979 ISBN 90-277-0963-7; Pb 90-277-0964-5
134. M. Marković and G. Petrović (eds.), *Praxis.* Yugoslav Essays in the Philosophy and Methodology of the Social Sciences. [Boston Studies in the Philosophy of Science, Vol. XXXVI]. 1979 ISBN 90-277-0727-8; Pb 90-277-0968-8
135. R. Wójcicki, *Topics in the Formal Methodology of Empirical Sciences.* 1979 ISBN 90-277-1004-X
136. G. Radnitzky and G. Andersson (eds.), *The Structure and Development of Science.* [Boston Studies in the Philosophy of Science, Vol. LIX] 1979 ISBN 90-277-0994-7; Pb 90-277-0995-5
137. J. C. Webb, *Mechanism, Mentalism and Metamathematics.* An Essay on Finitism. 1980 ISBN 90-277-1046-5
138. D. F. Gustafson and B. L. Tapscott (eds.), *Body, Mind and Method.* Essays in Honor of Virgil C. Aldrich. 1979 ISBN 90-277-1013-9
139. L. Nowak, *The Structure of Idealization.* Towards a Systematic Interpretation of the Marxian Idea of Science. 1980 ISBN 90-277-1014-7
140. C. Perelman, *The New Rhetoric and the Humanities.* Essays on Rhetoric and Its Applications. 1979 ISBN 90-277-1018-X; Pb 90-277-1019-8
141. W. Rabinowicz, *Universalizability.* A Study in Morals and Metaphysics. 1979 ISBN 90-277-1020-2
142. C. Perelman, *Justice, Law and Argument.* Essays on Moral and Legal Reasoning. 1980 ISBN 90-277-1089-9; Pb 90-277-1090-2
143. S. Kanger and S. Öhman (eds.), *Philosophy and Grammar.* Papers on the Occasion of the Quincentennial of Uppsala University. 1981 ISBN 90-277-1091-0
144. T. Pawlowski, *Concept Formation in the Humanities and the Social Sciences.* 1980 ISBN 90-277-1096-1
145. J. Hintikka, D. Gruender and E. Agazzi (eds.), *Theory Change, Ancient Axiomatics and Galileo's Methodology.* Proceedings of the 1978 Pisa Conference on the History and Philosophy of Science, Volume I. 1981 ISBN 90-277-1126-7
146. J. Hintikka, D. Gruender and E. Agazzi (eds.), *Probabilistic Thinking, Thermodynamics, and the Interaction of the History and Philosophy of Science.* Proceedings of the 1978 Pisa Conference on the History and Philosophy of Science, Volume II. 1981 ISBN 90-277-1127-5
147. U. Mönnich (ed.), *Aspects of Philosophical Logic.* Some Logical Forays into Central Notions of Linguistics and Philosophy. 1981 ISBN 90-277-1201-8

148. D. M. Gabbay, *Semantical Investigations in Heyting's Intuitionistic Logic*. 1981 ISBN 90-277-1202-6
149. E. Agazzi (ed.), *Modern Logic – A Survey*. Historical, Philosophical, and Mathematical Aspects of Modern Logic and Its Applications. 1981 ISBN 90-277-1137-2
150. A. F. Parker-Rhodes, *The Theory of Indistinguishables*. A Search for Explanatory Principles below the Level of Physics. 1981 ISBN 90-277-1214-X
151. J. C. Pitt, *Pictures, Images, and Conceptual Change*. An Analysis of Wilfrid Sellars' Philosophy of Science. 1981 ISBN 90-277-1276-X; Pb 90-277-1277-8
152. R. Hilpinen (ed.), *New Studies in Deontic Logic*. Norms, Actions, and the Foundations of Ethics. 1981 ISBN 90-277-1278-6; Pb 90-277-1346-4
153. C. Dilworth, *Scientific Progress*. A Study Concerning the Nature of the Relation between Successive Scientific Theories. 2nd, rev. and augmented ed., 1986 ISBN 90-277-2215-3; Pb 90-277-2216-1
154. D. W. Smith and R. McIntyre, *Husserl and Intentionality*. A Study of Mind, Meaning, and Language. 1982 ISBN 90-277-1392-8; Pb 90-277-1730-3
155. R. J. Nelson, *The Logic of Mind*. 2nd. ed., 1989 ISBN 90-277-2819-4; Pb 90-277-2822-4
156. J. F. A. K. van Benthem, *The Logic of Time*. A Model-Theoretic Investigation into the Varieties of Temporal Ontology, and Temporal Discourse. 1983; 2nd ed., 1991 ISBN 0-7923-1081-0
157. R. Swinburne (ed.), *Space, Time and Causality*. 1983 ISBN 90-277-1437-1
158. E. T. Jaynes, *Papers on Probability, Statistics and Statistical Physics*. Ed. by R. D. Rozenkrantz. 1983 ISBN 90-277-1448-7; Pb (1989) 0-7923-0213-3
159. T. Chapman, *Time: A Philosophical Analysis*. 1982 ISBN 90-277-1465-7
160. E. N. Zalta, *Abstract Objects*. An Introduction to Axiomatic Metaphysics. 1983 ISBN 90-277-1474-6
161. S. Harding and M. B. Hintikka (eds.), *Discovering Reality*. Feminist Perspectives on Epistemology, Metaphysics, Methodology, and Philosophy of Science. 1983 ISBN 90-277-1496-7; Pb 90-277-1538-6
162. M. A. Stewart (ed.), *Law, Morality and Rights*. 1983 ISBN 90-277-1519-X
163. D. Mayr and G. Süssmann (eds.), *Space, Time, and Mechanics*. Basic Structures of a Physical Theory. 1983 ISBN 90-277-1525-4
164. D. Gabbay and F. Guenthner (eds.), *Handbook of Philosophical Logic*. Vol. I: Elements of Classical Logic. 1983 ISBN 90-277-1542-4
165. D. Gabbay and F. Guenthner (eds.), *Handbook of Philosophical Logic*. Vol. II: Extensions of Classical Logic. 1984 ISBN 90-277-1604-8
166. D. Gabbay and F. Guenthner (eds.), *Handbook of Philosophical Logic*. Vol. III: Alternative to Classical Logic. 1986 ISBN 90-277-1605-6
167. D. Gabbay and F. Guenthner (eds.), *Handbook of Philosophical Logic*. Vol. IV: Topics in the Philosophy of Language. 1989 ISBN 90-277-1606-4
168. A. J. I. Jones, *Communication and Meaning*. An Essay in Applied Modal Logic. 1983 ISBN 90-277-1543-2
169. M. Fitting, *Proof Methods for Modal and Intuitionistic Logics*. 1983 ISBN 90-277-1573-4
170. J. Margolis, *Culture and Cultural Entities*. Toward a New Unity of Science. 1984 ISBN 90-277-1574-2
171. R. Tuomela, *A Theory of Social Action*. 1984 ISBN 90-277-1703-6

172. J. J. E. Gracia, E. Rabossi, E. Villanueva and M. Dascal (eds.), *Philosophical Analysis in Latin America*. 1984 ISBN 90-277-1749-4
173. P. Ziff, *Epistemic Analysis*. A Coherence Theory of Knowledge. 1984 ISBN 90-277-1751-7
174. P. Ziff, *Antiaesthetics*. An Appreciation of the Cow with the Subtile Nose. 1984 ISBN 90-277-1773-7
175. W. Balzer, D. A. Pearce, and H.-J. Schmidt (eds.), *Reduction in Science*. Structure, Examples, Philosophical Problems. 1984 ISBN 90-277-1811-3
176. A. Peczenik, L. Lindahl and B. van Roermund (eds.), *Theory of Legal Science*. Proceedings of the Conference on Legal Theory and Philosophy of Science, Lund, Sweden, 11–14 December 1983. 1984 ISBN 90-277-1834-2
177. I. Niiniluoto, *Is Science Progressive?* 1984 ISBN 90-277-1835-0
178. B. K. Matilal and J. L. Shaw (eds.), *Analytical Philosophy in Comparative Perspective*. Exploratory Essays in Current Theories and Classical Indian Theories of Meaning and Reference. 1985 ISBN 90-277-1870-9
179. P. Kroes, *Time: Its Structure and Role in Physical Theories*. 1985 ISBN 90-277-1894-6
180. J. H. Fetzer, *Sociobiology and Epistemology*. 1985 ISBN 90-277-2005-3; Pb 90-277-2006-1
181. L. Haaparanta and J. Hintikka, *Frege Synthesized*. Essays on the Philosophical and Foundational Work of Gottlob Frege. 1986 ISBN 90-277-2126-2
182. M. Detlefsen, *Hilbert's Program*. An Essay on Mathematical Instrumentalism. 1986 ISBN 90-277-2151-3
183. J. L. Golden and J. J. Pilotta (eds.), *Practical Reasoning in Human Affairs*. Studies in Honor of Chaim Perelman. 1986 ISBN 90-277-2255-2
184. H. Zandvoort, *Models of Scientific Development and the Case of Nuclear Magnetic Resonance*. 1986 ISBN 90-277-2351-6
185. I. Niiniluoto, *Truthlikeness*. 1987 ISBN 90-277-2354-0
186. W. Balzer, C. U. Moulines and J. D. Sneed, *An Architectonic for Science*. The Structuralist Program. 1987 ISBN 90-277-2403-2
187. D. Pearce, *Roads to Commensurability*. 1987 ISBN 90-277-2414-8
188. L. M. Vaina, *Matters of Intelligence*. Conceptual Structures in Cognitive Neuroscience. 1987 ISBN 90-277-2460-1
189. H. Siegel, *Relativism Refuted*. A Critique of Contemporary Epistemological Relativism. 1987 ISBN 90-277-2469-5
190. W. Callebaut and R. Pinxten, *Evolutionary Epistemology*. A Multiparadigm Program, with a Complete Evolutionary Epistemology Bibliograph. 1987 ISBN 90-277-2582-9
191. J. Kmita, *Problems in Historical Epistemology*. 1988 ISBN 90-277-2199-8
192. J. H. Fetzer (ed.), *Probability and Causality*. Essays in Honor of Wesley C. Salmon. 1988 ISBN 90-277-2607-8; Pb 1-5560-8052-2
193. A. Donovan, L. Laudan and R. Laudan (eds.), *Scrutinizing Science*. Empirical Studies of Scientific Change. 1988 ISBN 90-277-2608-6
194. H.R. Otto and J.A. Tuedio (eds.), *Perspectives on Mind*. 1988 ISBN 90-277-2640-X
195. D. Batens and J.P. van Bendegem (eds.), *Theory and Experiment*. Recent Insights and New Perspectives on Their Relation. 1988 ISBN 90-277-2645-0

196. J. Österberg, *Self and Others*. A Study of Ethical Egoism. 1988 ISBN 90-277-2648-5
197. D.H. Helman (ed.), *Analogical Reasoning*. Perspectives of Artificial Intelligence, Cognitive Science, and Philosophy. 1988 ISBN 90-277-2711-2
198. J. Wolenski, *Logic and Philosophy in the Lvov-Warsaw School*. 1989 ISBN 90-277-2749-X
199. R. Wójjcicki, *Theory of Logical Calculi*. Basic Theory of Consequence Operations. 1988 ISBN 90-277-2785-6
200. J. Hintikka and M.B. Hintikka, *The Logic of Epistemology and the Epistemology of Logic*. Selected Essays. 1989 ISBN 0-7923-0040-8; Pb 0-7923-0041-6
201. E. Agazzi (ed.), *Probability in the Sciences*. 1988 ISBN 90-277-2808-9
202. M. Meyer (ed.), *From Metaphysics to Rhetoric*. 1989 ISBN 90-277-2814-3
203. R.L. Tieszen, *Mathematical Intuition*. Phenomenology and Mathematical Knowledge. 1989 ISBN 0-7923-0131-5
204. A. Melnick, *Space, Time, and Thought in Kant*. 1989 ISBN 0-7923-0135-8
205. D.W. Smith, *The Circle of Acquaintance*. Perception, Consciousness, and Empathy. 1989 ISBN 0-7923-0252-4
206. M.H. Salmon (ed.), *The Philosophy of Logical Mechanism*. Essays in Honor of Arthur W. Burks. With his Responses, and with a Bibliography of Burk's Work. 1990 ISBN 0-7923-0325-3
207. M. Kusch, *Language as Calculus vs. Language as Universal Medium*. A Study in Husserl, Heidegger, and Gadamer. 1989 ISBN 0-7923-0333-4
208. T.C. Meyering, *Historical Roots of Cognitive Science*. The Rise of a Cognitive Theory of Perception from Antiquity to the Nineteenth Century. 1989 ISBN 0-7923-0349-0
209. P. Kosso, *Observability and Observation in Physical Science*. 1989 ISBN 0-7923-0389-X
210. J. Kmita, *Essays on the Theory of Scientific Cognition*. 1990 ISBN 0-7923-0441-1
211. W. Sieg (ed.), *Acting and Reflecting*. The Interdisciplinary Turn in Philosophy. 1990 ISBN 0-7923-0512-4
212. J. Karpiński, *Causality in Sociological Research*. 1990 ISBN 0-7923-0546-9
213. H.A. Lewis (ed.), *Peter Geach: Philosophical Encounters*. 1991 ISBN 0-7923-0823-9
214. M. Ter Hark, *Beyond the Inner and the Outer*. Wittgenstein's Philosophy of Psychology. 1990 ISBN 0-7923-0850-6
215. M. Gosselin: *Nominalism and Contemporary Nominalism*. Ontological and Epistemological Implications of the Work of W.V.O. Quine and of N. Goodman. 1990 ISBN 0-7923-0904-9
216. J.H. Fetzer, D. Shatz and G. Schlesinger (eds.): *Definitions and Definability*. Philosophical Perspectives. 1991 ISBN 0-7923-1046-2

Previous volumes are still available.

KLUWER ACADEMIC PUBLISHERS – DORDRECHT / BOSTON / LONDON